Repair and Maintenance of Tractors and Power Tillers

CV CONTENT VIBES
PUBLISHERS • PRINTERS • AGGREGATORS

About the Author

Tejas Ashok Bhosale completed his Bachelor of Technology (B. Tech.) from Mahatma Phule Krishi Vidyapeeth (MPKV), Rahuri, Maharashtra. He qualified for the Junior Research Fellowship (JRF) examination conducted by the Indian Council of Agricultural Research (ICAR). Subsequently, he earned his Master of Technology (M. Tech.) from G.B. Pant University of Agriculture and Technology (GBPUAT), Pantnagar. He has also qualified for the ICAR Agricultural Scientists Recruitment Board National Eligibility Test (ICAR-ASRB-NET). With over 12 years of teaching experience in State Agricultural Universities (SAUs), Tejas is currently serving as an Assistant Professor at Veer Chandra Singh Garhwali Uttarakhand University of Horticulture and Forestry (VCSGUUHF), Bharsar, Pauri Garhwal, Uttarakhand. He is a DGCA-certified drone pilot, specializing in the application of drones within the agricultural sector. Tejas has published 7 research papers, 5 popular articles, 1 book chapter, and 1 book. He has handled one project as Principal Investigator (PI) and another as Co-Principal Investigator (Co-PI). Additionally, he possesses administrative experience in various capacities within SAUs. His areas of specialization include agricultural mechanization, precision agriculture, the use of drones in agriculture, and the design and development of intercultural and harvesting machinery. Tejas's expertise and contributions significantly enhance agricultural technology and education.

Repair and Maintenance of Tractors and Power Tillers

Tejas Ashok Bhosale
Assistant Professor
Agricultural Engineering
Department of Natural Resource Management
College of Horticulture
Veer Chandra Singh Garhwali Uttarakhand University of Horticulture and Forestry
Bharsar, Pauri Garhwal, Uttarakhand

CV CONTENT VIBES
PUBLISHERS • PRINTERS • AGGREGATORS

CONTENT VIBES
PUBLISHERS • PRINTERS • AGGREGATORS

1168, Sector 13, Urban Estate
Karnal-132 001, Haryana
Tel: 91-84470 75807, 18440 41168
Email: contentvibesppa@gmail.com
www.contentvibes.in

Print ISBN: 978-93-67550-39-7

ebooks ISBN: 978-93-67552-45-2

Preface

Agriculture is the backbone of the Indian economy, and mechanization has played a crucial role in improving farm productivity and efficiency. Tractors and power tillers have become indispensable for modern farming, reducing manual labor and enhancing operational efficiency. However, to ensure their long-term performance, regular repair and maintenance are essential. This book, Repair and Maintenance of Tractors and Power Tillers, is designed to provide comprehensive knowledge on the upkeep, troubleshooting, and servicing of these essential agricultural machines.

This book covers key aspects of tractor and power tiller maintenance, including routine servicing, fault diagnosis, engine overhauling, transmission repairs, fuel system troubleshooting, electrical system maintenance, and hydraulic system care. It also provides step-by-step guidance on preventive maintenance, safety measures, and sustainable servicing practices. The content is presented in a structured manner, making it useful for students, technicians farmers, and agricultural engineers.

Written with practical applications in mind, this book incorporates case studies, illustrations, and real-world examples to help readers understand maintenance procedures effectively. It also addresses modern technological advancements, such as electronic fuel injection systems, smart diagnostics and automation in farm machinery servicing. By following the maintenance guidelines provided in this book, users can enhance the efficiency, reliability and lifespan of their tractors and power tillers, ultimately contributing to improved agricultural output.

I hope this book serves as a valuable resource for farmers, mechanics students, and professionals in the agricultural machinery sector. Constructive feedback and suggestions for improvement are always welcome.

Author

Contents

1

Tractor and Power Tiller Introduction

INTRODUCTION

Tractors used for agricultural activities are used with many different agricultural implements for plowing, planting, harvesting and other operations. Tractors used for construction and industrial activities are used to pull and operate machinery used for transporting loads, digging, grading and other operations.

Tractors and power tillers are pivotal machines that have revolutionized agricultural and industrial practices across the globe. These versatile vehicles have played an essential role in enhancing productivity, efficiency, and speed in a wide range of operations. In agriculture, tractors are used for various tasks such as plowing, tilling, planting, irrigation, and harvesting. They serve as the backbone of modern farming, significantly reducing labor costs and the time required to complete essential activities. Coupled with specialized implements, tractors can perform tasks with remarkable precision and consistency, resulting in increased crop yields and improved soil health.

Power tillers, on the other hand, offer a more compact and affordable alternative to tractors, especially in regions where small-scale farming predominates. These machines are primarily used for tilling the soil, making them particularly valuable in cultivating land in small to medium-sized agricultural fields. Despite their smaller size and lower horsepower compared to tractors, power tillers are capable of performing various tasks such as weeding, leveling, and inter-row cultivation, providing farmers with a cost-effective solution for enhancing farm productivity.

Both tractors and power tillers are indispensable in modern agriculture, as they provide the necessary power and efficiency to carry out a wide range of field operations. Moreover, these machines have found applications in construction, road development, and other industries, where they are used for tasks like hauling, digging, and leveling. With technological advancements, the capabilities of tractors and power tillers continue to evolve, enabling them to meet the growing demands of both the agricultural and industrial sectors. This chapter will explore the features, types, applications, and maintenance practices related to tractors and power tillers, highlighting their significance in shaping the future of modern agriculture and industry.Tractors play an important role in the agricultural and industrial sectors, making them indispensable vehicles in the modern world.

TRACTORS

A tractor is a self-propelled power unit equipped with either wheels or tracks, designed to operate agricultural tools, machines, and trailers. The tractor's engine serves as the prime mover for active tools and stationary farm machinery, typically through a power take-off (PTO) or belt pulley system.

Figure: 1.1 Wheel Tractors

Tractors have played a vital role in the advancement of agriculture and food production, proving to be an indispensable machine that has continually enhanced our lives by saving time and effort. These machines are essential to the agricultural sector, enabling the efficient operation of farms, even on land that would otherwise be difficult to cultivate. Tractors also help reduce fuel expenses, further contributing to their importance in modern farming practices.

Various of Tractors:

Garden Tractors:

Garden tractors are highly beneficial for gardening, simplifying the entire process and helping save both time and money. These tractors are excellent tools for gardeners, as they significantly speed up tasks and increase efficiency, making gardening work much easier and more productive.

Implement carrier tractors:

Implement carrier tractors are an ideal solution for farmers who need to transport their crops from the field to the market. They reduce the need for manual labor and offer a convenient method of transporting crops across different locations. These tractors are specifically designed for use on highways, built to carry heavy loads at high speeds, making them an efficient tool for long-distance transportation.

Advantages of Tractors

- **Increased Efficiency:** Tractors improve farm productivity by reducing the time required for various agricultural tasks.
- **Versatility:** Tractors can be used with various implements for tasks like plowing, planting, and harvesting.
- **Reduced Labor Requirement:** Tractors reduce the need for manual labor, allowing farmers to do more work in less time.
- **Improved Land Utilization:** Tractors can operate on various types of land, including those unsuitable for traditional farming.
- **Higher Speed and Precision:** Tractors perform tasks more quickly and with greater precision than manual labor.
- **Increased Yield:** The ability to cover large areas quickly increases the overall agricultural yield.
- **Capability for Heavy Tasks:** Tractors are capable of performing heavy tasks like towing large loads and digging.

Disadvantages of Tractors

- **High Initial Cost:** The purchase cost of tractors can be expensive, especially for small-scale farmers.
- **Fuel Consumption:** Tractors often consume a lot of fuel, leading to higher operational costs.
- **Maintenance Costs:** Tractors require regular maintenance, and parts can be costly to replace.
- **Environmental Impact:** The use of tractors can contribute to soil compaction and environmental degradation if not used properly.
- **Dependence on Fuel:** Tractors are dependent on fuel, which can be a limitation during shortages or price hikes.

- **Noise and Pollution:** Tractors generate noise and air pollution, contributing to environmental pollution.
- **Training Requirement:** Proper training is needed to operate tractors effectively and safely.

TIME LINE: TRACTOR DEVELOPMENT

The modern tractor is the product of a gradual evolution of machinery over several stages. Below is the time line of tractor development:

- **1890:** The term "tractor" was first recorded in a patent issued for a tractor or traction engine invented by George H. Harris of Chicago.
- **1906 :** Charles W. Hart and Charles H. Parr from Charles City, Iowa, introduced the first successful gasoline tractor.
- **1908:** The first Winnipeg tractor trials were held.
- **1911:** The first tractor demonstration took place in Omaha, Nebraska.
- **1915-1919:** The power take-off (PTO) system was introduced.
- **1920-1924:** The all-purpose tractor was developed.
- **1936-1937:** Diesel engines were used in tractors, and pneumatic tires were introduced.
- **1950-1960:** Diesel tractor manufacturing expanded globally on a large scale.
- **1960-1961:** Tractor manufacturing began in India, with M/s Eicher Good Earth being the first manufacturer.
- **1962-1970:** Manufacturers like Tractor and Farm Equipment in Madras, Hindustan Tractors in Baroda, Escorts Tractors in Faridabad, and International Harvester in Bombay began production during this period.
- **1971:** Escorts Tractor Ltd. started producing Ford Tractors.
- **1973:** The production of HMT Tractors commenced.
- **1974:** The manufacture of Pitti and Kirlosker Tractors began.
- **1975:** Harsha Tractors was established.
- **1981:** Auto tractors began production.
- **1982:** Universal Tractors was founded.
- **1983-2003:** During this period, tractors from brands such as GTCL, M.M Tractors, Sonalika, VST, L&T, and Bajaj Tractors were produced.

CLASSIFICATION OF TRACTORS

Tractors can be classified into three classes on the basis of structural design:

1. Wheel tractor
2. Crawler tractor
3. Walking type tractor

Wheel Tractor

Tractors having three or four pneumatic wheels are called wheel tractors. Four wheel tractors are popular everywhere

Figure: 1.2 Three Wheel Tractor

Crawler Tractor

Inother words it is also called Track type tractor or Chain type tractor. In such tractors , there is endless chain or track in place of pneumatic wheels

Figure: 1.3 Crawler Tractors

Walking Type or Power Tiller

Power tiller is a walking type tractor. This tractor is usually fitted with two wheels only. The direction of travel and its controls for field operation is performed by the operator, walking behind the tractor

Figure: 1.4 Power Tillers

Classification of Wheeled Tractors

On the basis of purpose, wheeled tractors are classified into three groups:

- General purpose tract
- Row crop tractor
- Special purpose tractor

General Purpose Tractor

It is used for major farm operations such as ploughing, sowing, harvesting and transporting works. Such tractors have i) low ground clearance ii)Increased engine power iii) good adhesion and iv) wide tyres

Figure: 1.5 General Purpose Tractors

Row Crop Tractors

It is used for row crop cultivation. Such tractor is provided with replaceable driving wheels of different tread widths. It has high ground clearance to save damage of crops. Wide wheel track can be adjusted to suit inter row distance.

Figure: 1.6 Row Crop Tractors

Special Purpose Tractor

It is used for definite jobs like cotton fields, marshy lands, hill sides, garden etc. special designs are there for special purpose tractor.

For example. b) Tractor with winch unit a) multi drive tractor c) tractor for golf grounds etc.

Figure: 1.7 (a) Multi Drive Tractors

Figure: 1.7 (b) Tractor with Winch Unit

Figure: 1.7 (c) Tractor for Golf Grounds

SELECTION OF TRACTOR

The selection of a tractor depends on several factors:

- **Land Holding:** For a single cropping pattern, it is generally recommended to allocate 1 hp for every 2 hectares of land. Therefore, a tractor with 20-25 hp would be suitable for a 40-hectare farm.

- **Cropping Pattern:** When adequate irrigation is available and multiple crops are grown, a recommendation of 1.5 hectares per hp is common. For such conditions, a 30-35 hp tractor would be appropriate for 40 hectares of land.
- **Soil Condition:** The tractor's suitability also depends on soil type. A tractor with a shorter wheelbase, higher ground clearance, and lower weight is better for lighter soils, but may not provide enough depth in heavier soils like black cotton soil.
- **Climatic Conditions:** In very hot or desert areas, air-cooled engines are preferable due to their better heat dissipation, whereas for higher altitudes, air-cooled engines are also favored as water-cooled engines can freeze in cold temperatures.
- **Repair Facilities:** It is important to ensure that the tractor has an accessible dealer nearby, offering necessary repair and maintenance services with adequate technical expertise.
- **Running Cost:** Tractors with lower specific fuel consumption should be prioritized to reduce overall running costs.
- **Initial Cost and Resale Value:** While considering the resale value, the initial purchase cost should not be excessively high to avoid substantial interest payments.
- **Test Report:** It's advisable to consult test reports from farm machinery testing stations to get informed guidance on the performance and reliability of the tractor.

COMPONENTS OF WHEEL TRACTOR

These are the important components of wheel tractor:

1. Final drive
2. Real wheel
3. Front wheels
4. Steering mechanism
5. I.C Engine
6. Clutch
7. Transmission gears
8. Differential unit
9. Tractor pulley
10. Draw bar and
11. Control panel
12. Hydraulic control and hitch system
13. Brakes
14. Power-take-off unit

OPERATION OF A TRACTOR

The operation of a tractor is based on the principle of transferring engine power to the wheels or tracks via power transmission components. The engine, typically a diesel engine, is at the core of a tractor. It generates mechanical motion by converting the energy produced from compressing and igniting a fuel-air mixture. This motion is then transmitted from the engine's output shaft to the power transmission system.

Tractor operation is managed through the control console within the driver's cabin. The console includes essential components such as the accelerator pedal, brake pedal, clutch pedal, steering wheel, gear lever, and various other control switches.

Power transmission components in tractors include a hydraulic pump, clutch, transmission, rear axle, front axle, and differential. These components work together to transfer power and motion to the wheels or tracks of the tractor.

The tractor's movement is primarily driven by the rear wheels or tracks, with power being used to rotate them via the differential. The rear axle, equipped with attachments like a tow ring or tow bar, enables the tractor to tow other vehicles or move machinery.

In short, tractors rely on their robust engines and intricate power transmission components to move, and they serve crucial roles across agriculture, construction, and industrial fields.

ENGINE STRUCTURE OF TRACTORS

Tractors are primarily equipped with diesel engines that generate mechanical power through the compression and combustion of fuel. These engines are built with a complex structure to ensure high performance and durability, which are essential for the demanding tasks tractors perform.

The design of tractor engines emphasizes power and longevity. Key components of the engine include the cylinder block, cylinder head, pistons, crankshaft, camshaft, fuel system, air system, lubrication system, and cooling system.

The cylinder block serves as the main frame of the engine, housing the cylinders, which typically range from 4 to 12 in number. The pistons move within these cylinders, compressing the combustion chamber to generate power. The cylinder head seals the top of the cylinders and provides a connection to the cylinder head gasket.

As the pistons move inside the cylinders, they compress the combustion chamber and transfer energy via connecting rods linked to the crankshaft. The crankshaft then converts this linear motion into rotational motion, transmitting power to the transmission system.

The fuel system of the tractor ensures the engine's fuel requirements are met by drawing fuel from the tank and delivering it to the combustion chambers through a fuel pump. Similarly, the air system facilitates the engine's air intake by pulling clean air through an air filter and directing it into the cylinders.

To minimize wear and friction, the lubrication system pumps clean oil to various engine components via an oil filter. Meanwhile, the cooling system regulates engine temperature by circulating coolant or air to prevent overheating.

POWER TILLER

A power tiller is a compact agricultural machine designed primarily for soil tilling, to prepare the ground for crop planting. It is a versatile and efficient tool, particularly suited for small-scale farming or regions where access to larger tractors is limited. Power tillers are typically used for various tasks such as plowing, harrowing, and leveling the soil. They are smaller, lighter, and more maneuverable than traditional tractors, making them ideal for working in smaller or confined spaces.

These machines are powered by small engines, typically ranging from one to ten horsepower, and can run on gasoline, petrol, or a mixture of oil. The design allows the operator to control the movement by adjusting the handlebars, which helps manage friction as the machine moves gently forward. Most power tillers are designed to move forward only and cannot reverse, a feature incorporated for safety.

A power tiller is a type of prime mover where the operator walks behind the machine to control its direction and operation in the field. Also known as a hand tractor or walking-type tractor, the concept of the power tiller was first introduced globally in the 1920s. Japan was the first country to use power tillers on a large scale, with the first successful model developed in 1947. India adopted power tillers in 1963, and since then, the manufacturing of various models, including those from Iseki, Sato, Krishi, Kubota, Yanmar, and Mitsubishi, has been initiated in India.

Figure: 1.8 In agriculture, power tillers are indispensable for plowing, sowing, spraying, harvesting, and transporting tasks. They are particularly valuable in rice cultivation, where they are commonly used for puddling operations, making them one of the most sought-after machines for this purpose.

Advantages of Power tiller tractors

- Power tiller Tractors are a kind of tractor that can turn the soil to make it easier for planting.
- Power tiller Tractors are not just tools for farmers but can also be used in gardens and landscaping.

- They help prepare the ground before planting by loosening soil, levelling it, and creating pathways.
- Power tiller Tractors are farm equipment that helps in the cultivation process. They are a hybrid between a tractor and an implement.
- Higher productivity with less effort
- Less fuel consumption
- Improved soil quality

Disadvantages of Power Tiller Tractors

- Power tiller Tractors are more expensive and less efficient than traditional tractors. They also have a shorter lifespan.
- Power tiller Tractors are designed for a specific purpose – to till and cultivate the land. They are not designed for other activities such as mowing lawns, cutting grass, and trimming hedges.
- Power tiller Tractors have many disadvantages when compared to traditional garden tractors. They are more expensive and difficult to maintain. They also require much fuel and cannot be used in soft soil or mud areas.
- A power tiller is not easy to handle and requires much skill to drive safely.

IMPORTANCE OF POWER TILLER TRACTORS

Power tiller tractors are crucial tools for farmers, primarily used for tilling the soil and getting it ready for planting. These machines feature a tiller attachment that is essential for cultivating the land. Power tillers are typically large, robust machines with extensive tracks that provide optimal traction, even on soft or uneven ground. The significance of power tiller tractors extends beyond farming, as they are also valuable to landscapers and construction companies who rely on them for various tasks. These tractors are versatile, assisting with several agricultural activities, including plowing fields for seed planting, harvesting crops, fertilizing the soil, and preparing it for cultivation.

COMPONENTS OF POWER TILLER

These are the components of a power tiller:

1. Engine
2. Transmission gears
3. Clutch
4. Brakes
5. Rotary unit

All power tillers are fitted with an I.C. engine. At present makes like Kubota, Mitsubishi, Krishi, Yanmar and Satoh use diesel engine. Iseki make use kerosene engine.

OPERATION OF A POWER TILLER

For operation of power tiller, the power is obtained from the I>C engine, fitted on the power tiller. The engine power goes to the main clutch with the help of belt or chain. From main clutch power is divided in two routes, one goes to transmission geras, steering clutch and the to wheels. The other component goes to the tilling clutch and then to tilling attachment .V- belt pulley is usually used to transmit power from the engine to the main clutch

- **Main clutch:** Power goes from engine to main clutch through V- belt – pulley arrangement
- **Transmission gears:** Transmission gears consists of gears, shafts, and bearings. Transmission gears reduce speed of the engine and increase the torque at the wheels.
- **Brakes:** All power tillers have braking arrangement for stopping the movement of power tiller.
- **Wheels:** Usually two to four ply pneumatic tyre are used in power tillers for the wheels. The inflation pressure ranges from 1.1 to 1.4 kg/cm^2.
- **Rotary unit:** Power tillers have a rotary unit for field operation. Rotary tines are used in rotary unit for soil cutting and pulverizing purposes. Rotary tines are of three types:
    - Straight tines
    - Curved tines
    - L shape blades

Engine System

In india, most power tillers utilize a single-cylinder, 4-stroke, air-cooled diesel with a small horsepower range. It is designed to be fule-efficient and suitable for small-seals farming operations. The key point about the engine system of a typical indian power tiller:

- **Type of Engine:** Power tillers are generally powered by internal combustion engines, with most using diesel engines. However, gasoline engines are also used in smaller models.
    - **Horsepower:** Power tillers typically have engine power ranging from 5 HP to 15 HP, depending on the model.
- **Cooling System:** The engine may use either air-cooled or liquid-cooled systems. Air-cooled engines are lighter and simpler, but liquid-cooled engines provide better performance in hot conditions.

Fuel System: Fuel is stored in a fuel tank, and fuel injectors or carburetors deliver the appropriate amount of fuel to the engine for combustion.

Transmission System

- **Gearbox:** Power tillers have a transmission system that allows the operator to select different gears to adjust speed. This allows for varying the tilling speed based on the soil's hardness and the task at hand.
- **Drive System:** Most power tillers have a chain drive or belt drive system, which transmits power from the engine to the wheels and tilling blades.

- **Forward and Reverse Gear:** Power tillers typically offer forward and reverse gear options, making it easier to maneuver in tight spaces or when turning.

Tilling Mechanism

- **Rotary Blades/Drums:** The tiller's primary function is to break up the soil, and this is achieved using rotary blades or tilling drums. These blades rotate and break up the soil, making it looser for planting.
    - The blades can be adjusted based on the soil type and tilling depth.
    - Some models have adjustable blade systems, where the angle or width of the blades can be modified to suit the work.
- **Tine System:** The rotary tiller often uses tines (sharp, twisted metal blades) to break the soil. The tines can vary in shape, length, and design, depending on whether the tiller is used for lighter or more intense tilling.

Steering and Drive Wheels

- **Drive Wheels:** Power tillers have two or four wheels that provide traction. These wheels are powered by the engine, allowing the machine to move forward and backward.
- **Steering:** Steering can be either manual or powered, with larger tillers offering more sophisticated steering mechanisms for easier operation.
- **Handlebars:** The tiller has handlebars for the operator to control the machine. These may be adjustable for comfort and to control the direction of movement.

Chassis and Frame

- **Frame:** The chassis or frame is made from durable materials like steel to ensure the tiller can handle rough agricultural work. The frame supports the engine, transmission, tilling equipment, and the operator's controls.
- **Weight Distribution:** Power tillers are designed with a balance between the weight of the engine, tilling components, and the wheels, ensuring proper stability when working in fields.

Clutch System

The clutch engages or disengages the engine from the rotary blades or wheels, allowing the operator to control when the blades are active. The clutch is crucial for safety and operational control, as it allows the operator to stop the tiller's movement without stopping the engine.

Lubrication and Cooling Systems

- **Oil Pump:** A lubrication system ensures that engine parts, especially the gearbox and rotary components, are adequately lubricated to prevent friction and wear.
- **Cooling System:** For engines, especially diesel engines, a cooling system (air or liquid) ensures that the engine operates within safe temperature ranges to prevent overheating during long hours of use.

Safety Features

- **Emergency Shut-Off:** Power tillers are equipped with an emergency stop system that allows the operator to quickly turn off the engine if necessary.
- **Safety Guards:** Blades and moving parts are often shielded with safety guards to prevent accidental contact.
- **Handbrake or Parking Brake:** A handbrake or parking brake is used to secure the tiller when it is not in operation, especially when working on sloped terrain.

Exhaust System

- **Exhaust Pipe:** Like all combustion engines, power tillers have an exhaust system to expel gases produced during engine operation. Some systems are designed to minimize emissions for environmental concerns.

TYPES OF POWER TILLERS

Power tillers come in various designs, and they are often categorized based on their size, engine type, and intended use.

Hand-Tiller or Walking Tiller

These are smaller, lighter machines designed for use in confined spaces or small fields. They are manually operated by walking behind the machine.

Riding Power Tiller

These are larger, more powerful models where the operator rides on the tiller. Riding models are often used for larger plots of land and offer greater stability and power.

Two-Wheel and Four-Wheel Power Tillers

- **Two-Wheel Power Tillers:** Most common for small-scale operations, these tillers have two wheels and are usually lighter and more maneuverable.
- **Four-Wheel Power Tillers:** Larger models with four wheels, offering enhanced stability and higher power for more intensive work.

APPLICATIONS OF POWER TILLERS

- **Soil Tilling:** The primary use of power tillers is for tilling soil to prepare fields for sowing crops.
- **Weed Control:** By breaking up the soil, power tillers also help control weeds by uprooting them or burying them under the soil.
- **Fertilizer and Compost Mixing:** Some power tillers can be equipped with attachments to mix organic matter, compost, and fertilizers into the soil.
- **Soil Aeration:** Power tillers help aerate the soil, which improves water infiltration and root growth.

- **Gardens and Orchards:** Smaller tillers are used in home gardening or orchards to maintain soil health.

A power tiller is a valuable tool for small-scale and medium-scale farmers. Its compact design, efficiency, and versatility make it ideal for working on smaller farms or plots where larger tractors cannot operate efficiently. Understanding the components and systems of a power tiller is essential for its proper maintenance and safe operation.

2

Components of a Tractor Engine

INTRODUCTION

The engine is the core of a tractor, providing the essential power required for all its functions. Typically, a tractor engine is a four-stroke internal combustion engine that operates on diesel, although some models may use gasoline. The engine consists of several integral components that work together to generate power and ensure smooth operation. These parts include the cylinder block, which houses the cylinders where fuel combustion occurs; the pistons, which move within the cylinders to convert the energy from combustion into mechanical motion; and the crankshaft, which transfers the piston's motion to the transmission. The valve system regulates the intake of air and fuel, as well as the exhaust of gases, ensuring efficient combustion. The fuel system, made up of the fuel tank, pump, and injectors, supplies fuel to the engine, while the cooling system, typically water-based, prevents the engine from overheating. Furthermore, the lubrication system ensures proper lubrication of all moving components to reduce friction and wear. The starter motor and battery are responsible for initiating the engine's operation, and the exhaust system directs exhaust gases away from the engine. Together, these components enable the tractor to generate the power needed for a wide range of agricultural and industrial tasks.

A tractor engine is designed to deliver high torque at low speeds. It is an internal combustion engine that primarily operates on diesel, compressing and burning the fuel to generate mechanical motion. Due to the engine's complex structure, it is composed of several key components. The following list outlines the major parts that make up a tractor engine:

Piston

The piston is a crucial component of an engine. It is a cylindrical shaft, typically lubricated, that fits into the opening of the engine's cylinder. The piston moves back and forth due to the pressure exerted by a fluid or gas. In an internal combustion (IC) engine, the piston moves as a result of the expanding gases within the cylinder head. It is connected to a pivoted connecting rod, which is linked to the crankshaft or flywheel. The primary function of the piston is to convert the reciprocating motion into rotational motion.

Figure: 2.1 Piston

A cylinder houses a piston that moves reciprocally to compress the fuel-air mixture and initiate combustion. The piston generates pressure through fuel combustion, which is transmitted to rotate the crankshaft via a connecting rod. The top part of the piston is called the crown, and the sides are referred to as the skirt. Pistons are designed to endure extreme temperatures and high pressures, which is why they are typically made from materials like cast iron or aluminum alloy. Piston rings are used to create a seal between the piston and the cylinder to prevent leakage.

The piston is a cylindrical component that is closed at one end and open at the other, ensuring a close sliding fit within the engine cylinder. The piston is connected to the connecting rod through a piston pin. When the expanding gases exert force against the closed end of the piston, the piston moves downward within the cylinder. This motion causes the connecting rod to rotate the crankshaft. The connecting rod is typically made from cast iron due to its high compressive strength, low coefficient of expansion, resistance to high temperatures, ease of casting, and affordability. Aluminum and its alloys are preferred for pistons primarily because of their lightweight nature and efficient heat dissipation properties.

CYLINDER

A cylinder is a key component of an engine, acting as the chamber where fuel combustion occurs to generate power. The more cylinders present, the more pistons are engaged in fuel combustion, leading to increased power output. Since cylinders endure the pressure and heat generated during combustion, they are made from strong materials like cast iron for durability. Additionally, a cooling system is essential to manage high temperatures and maintain operational efficiency.

- **3-cylinder Engines:** These engines are compact, fuel-efficient, and cost-effective, making them ideal for light to medium-duty tasks.
- **4-cylinder Engines:** Known for their power and versatility, these engines provide a balanced performance, making them suitable for medium to heavy-duty agricultural operations.
- **Considerations:** When choosing an engine, factors such as power requirements, fuel efficiency, budget, maintenance needs, and the tasks to be performed must be taken into account.

The engine is the heart of a powerful tractor, much like the soul of a human body. Thus, understanding the engine specifications is crucial when evaluating tractors. One of the important aspects to consider is the number of cylinders in the engine.

In the tractor industry, the two most common engine configurations are the three-cylinder and four-cylinder engines. Each configuration offers unique advantages and disadvantages, and a detailed comparison of 3-cylinder versus 4-cylinder engines can provide insights into which option suits specific needs and preferences.

Figure: 2.2 3-Cylinder Engine

3-Cylinder Engine

A 3-cylinder engine is a type of internal combustion engine distinguished by having three cylinders arranged in a straight line. These cylinders are aligned sequentially and share a single crankshaft. In contrast to the 4-cylinder engine, a 3-cylinder engine's firing order occurs at intervals of 120 degrees.

Benefits of 3-Cylinder Engine

- **Low Friction:** With less metal-to-metal contact, 3-cylinder engines experience lower friction losses, which boosts their overall operational efficiency.
- **Compact Size:** The compact design of the 3-cylinder engine allows it to fit into smaller spaces, making it suitable for a wide range of tractor configurations, especially those with limited space.
- **Higher Fuel Efficiency:** The streamlined design of 3-cylinder engines, with fewer moving parts, reduces friction and enhances fuel efficiency, making them more economical compared to 4-cylinder engines.
- **Cost-Effective:** Due to fewer components and less material required in their construction, 3-cylinder engines are generally more affordable to manufacture, resulting in lower overall costs for tractors equipped with these engines.

Disadvantages of 3-Cylinder Engine

- **Shorter Lifespan:** The workload placed on the three pistons tends to generate more stress, which may reduce the engine's overall lifespan when compared to engines with more pistons.
- **Limited Power:** With their smaller size and capacity, 3-cylinder engines are less suited for heavy-duty agricultural tasks, offering limited power output, which may not be sufficient for larger-scale operations.
- **Increased Vibration:** The task of balancing three pistons in a 3-cylinder engine can be difficult, leading to higher vibration levels when compared to 4-cylinder engines.
- **Noise Levels:** 3-cylinder engines typically produce more noise due to their lighter build and reduced space for noise-dampening components, making them noisier than their 4-cylinder counterparts.

Applications of 3-Cylinder Engines

Despite their drawbacks, 3-cylinder engines are widely used in various agricultural and industrial applications. Their compact size, fuel efficiency, and affordability make them ideal for small to medium-sized tractors used for tasks such as landscaping, gardening, and light agricultural work. Furthermore, their cost-effectiveness makes them an attractive choice for farmers and businesses that are budget-conscious but still require reliable, efficient machinery.

4-Cylinder Engine

A 4-cylinder engine is a type of internal combustion engines, that consists four separate cylinders, each with a seprate piston. The cylinders are generally organized in a straight line or a "V" shape, with each piston undertaking specific functions in the cycle of combustion.

Figure: 2.3 4-Cylinder Engine

Advantages of 4-Cylinder Engine

- **More Torque:** Generally, 4-cylinder engines produce more torque compared to 3-cylinder engines, which is vital for agricultural and industrial operations. Torque is the rotational force that allows the engine to pull heavier loads, making 4-cylinder engines ideal for handling heavy equipment or performing tough tasks. Higher torque ensures that the tractor can power through challenging terrains, heavy soil, or larger attachments without struggling.
- **Smooth Operation:** Due to the even firing order and additional cylinder, 4-cylinder engines typically produce less vibration, leading to smoother operation. This results in greater comfort for the operator, reducing physical strain and fatigue during long working hours. The smoother operation also reduces wear and tear on the engine and other mechanical components, contributing to the longevity of the vehicle. Additionally, reduced vibrations improve the overall stability of the vehicle, particularly when operating in uneven terrain or during high-speed travel.
- **Uniform Power Delivery:** The firing sequence in 4-cylinder engines ensures a balanced and consistent power output across a wide range of engine RPMs. This leads to smooth and stable performance, regardless of the load or operating conditions. The even distribution of power also prevents sudden power surges or dips, making these engines ideal for tasks that require steady performance.
- **Enhanced Response:** With four pistons constantly working, 4-cylinder engines offer superior responsiveness. Each piston takes turns in producing power, meaning that there's always a piston generating power in the engine cycle. This continuous generation of power results in better acceleration and a more immediate response, especially during sudden changes in speed or load.
- **Greater Power and Performance:** The presence of four cylinders enables a higher power output, allowing the engine to perform more efficiently across various tasks. A 4-cylinder engine can handle demanding operations such as deep plowing, large-scale tilling, or heavy hauling with ease. The multiple cylinders ensure that the engine doesn't have to work as hard for each task, making it more versatile and capable of handling a broader range of applications.

- **Improved Efficiency:** The higher number of cylinders allows for better heat distribution and more efficient fuel combustion. The engine can operate at optimal temperatures with less strain on individual parts, leading to improved efficiency in both power generation and fuel usage, ultimately resulting in longer engine life and reduced operational disruptions.

Disadvantages of 4-Cylinder Engine

- **Higher Cost:** The increased complexity of a 4-cylinder engine often results in higher initial costs compared to 3-cylinder engines. The additional components, such as extra pistons, cylinders, and the intricate valve system, contribute to the higher manufacturing cost. As a result, tractors equipped with 4-cylinder engines tend to have a higher purchase price, which can be a factor for budget-conscious buyers.
- **Increased Fuel Consumption:** With an additional cylinder, the 4-cylinder engine requires more fuel to generate power. This leads to increased fuel consumption, making it less fuel-efficient than smaller engines, such as the 3-cylinder type. Over time, this can increase the overall operating costs, especially for farmers or businesses with tight budgets. However, this disadvantage may be offset by the higher power and torque output, depending on the work requirements.
- **Higher Maintenance Costs:** Due to the increased number of components and complexity, 4-cylinder engines typically come with higher maintenance costs. More parts mean more potential for wear and tear, requiring regular checks and replacements. For example, maintenance of valves, pistons, and cylinder heads in a 4-cylinder engine is generally more labor-intensive and costly compared to 3-cylinder engines. The higher number of components also increases the likelihood of breakdowns, which can result in costly repairs if not properly maintained.

Applications of 4-Cylinder Engines

Despite their higher costs and fuel consumption, 4-cylinder engines are highly sought after in agricultural and industrial settings where power, efficiency, and versatility are essential. These engines are well-suited for medium to large-scale farming operations, including tasks such as plowing, tilling, sowing, and harvesting. Their consistent power delivery ensures that they can handle heavy workloads without compromise, making them invaluable for operations that require sustained output over long periods.

Moreover, the higher torque and smoother operation of 4-cylinder engines make them perfect for demanding tasks such as hauling, lifting, and towing heavy equipment. These engines are also commonly found in tractors and machinery used for road work, construction, and other industries where reliable performance is critical. Their ability to perform consistently at higher speeds or under tough conditions makes them an indispensable tool in various industries requiring power and reliability. Additionally, their ability to maintain efficiency and comfort during long workdays contributes to their preference in settings that demand continuous operation.

RINGS

Todays a piston ring assembly typically consists of two compression rings and one oil-control ring. The primary role of the top compression ring is to seal the combustion gases within the cylinder, preventing gas leakage and ensuring efficient power generation. The oil-control ring, situated at the bottom of the assembly , is responsible for regulating the amount of oil that coats the cylinder walls, preventing excessive oil consumption and minimizing smoke and emissions. The second compression ring, positioned between the top ring and the oil-control ring, generally serves a dual purpose, assisting in sealing the combustion gases while also contributing to oil control. This configuration helps maintain optimal engine performance by effectively regulating gas pressure and oil levels within the cylinder.

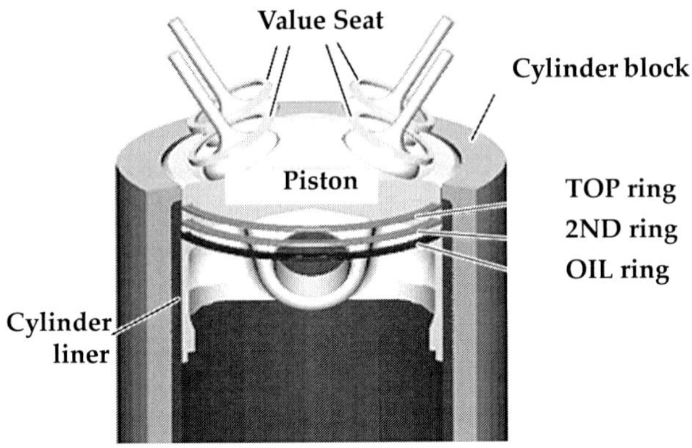

Figure: 2.4 Piston Ring Assembly

Tractor piston rings are crucial components of an internal combustion engine. They are metal rings that fit into grooves around the piston.

Figure: 2.5 Structure of Piston Rings

Together, these three rings work in unison to provide adequate lubrication within the cylinder bore, preventing excessive wear and ensuring the engine runs smoothly over time. Though small, piston rings are vital components that fit around the piston to form a seal between it and the cylinder wall. This seal maintains pressure within the combustion chamber, controls oil consumption, and helps cool the engine. Tractor piston rings are typically constructed from high-strength materials to endure the extreme temperatures and pressures that occur during engine operation.

Functions of Piston Rings

Sealing the Combustion Chamber

Piston rings establish a seal between the piston and the cylinder wall, preventing the leakage of combustion gases.Its function is to fully maintain in-cylinder airtightness between the piston and cylinder wall.

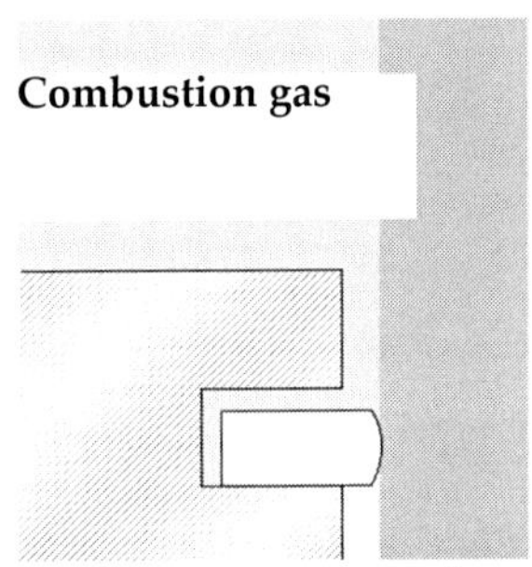

Figure: 2.6 Combustion Chamber

It works hard to keep a lid on combustion gas leaking into the gap between the piston and cylinder created at the time of an explosion. If combustion gas leaked, full power could not be realized and fuel consumption would increase, which is detrimental both economically and environmentally.

Controlling Oil Consumption

Piston rings control the amount of engine oil that reaches the combustion chamber, preventing excessive oil consumption and smoke production.

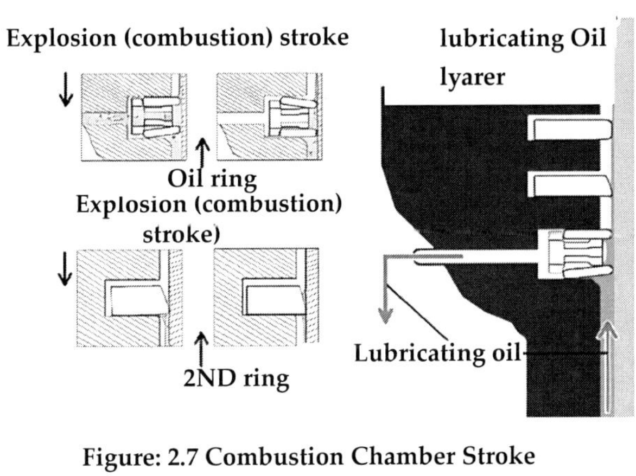

Figure: 2.7 Combustion Chamber Stroke

A fundamental function of piston rings is to requlate oil flow to the cylinder walls. The oil rings remove surplus oil off the cylinder walls and direct it back into the oil sump, preventing it from entering the combustion chamber and causing excessive smoke or oil burning.

Heat Dissipation

Piston rings facilitate the mission of heat from the piston to the cylinder wall, contributing to engine cooling.

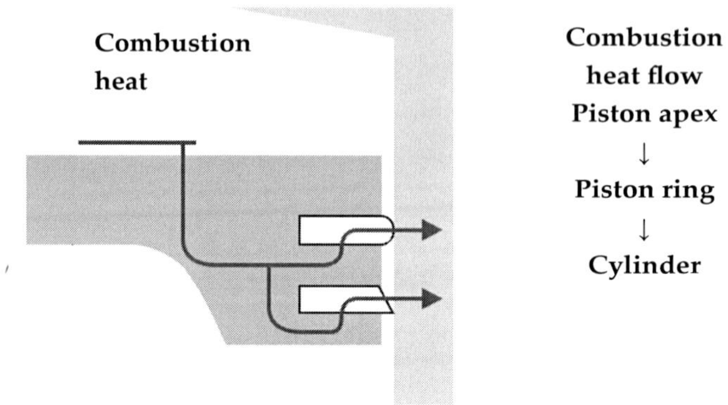

Figure: 2.8 Combustion Ineat Flow

Engines produce significant heat during operation, and the piston rings help in transferring this heat from the piston to the cylinder walls. This interaction keeps up with ideal engine temperature and forestalls overheating.

Maintaining Engine Compression:

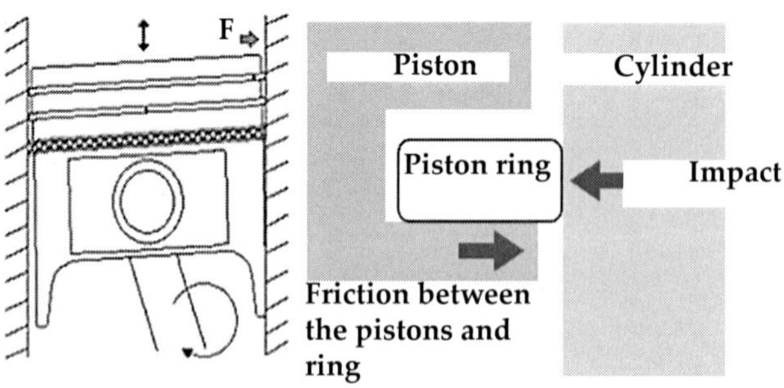

Figure: 2.9 Impact of Piston Ring

Optimcompression is essentional for efficient combustion. Piston rings help maintain the required compression in the engine, improving overall performance, fuel efficiency, and power output.

Types of Piston Rings

Compression Rings

Located at the top of the piston, compression rings are responsible for preventing combustion gases from escaping into the crankcase. These rings help maintain the pressure within the combustion chamber, ensuring optimal power production.

Oil Rings

Situated just below the compression rings, oil rings regulate the amount of oil that lubricates the cylinder walls. They play a vital role in minimizing oil consumption and keeping the engine clean by preventing excess oil from entering the combustion area.

Scraper Rings

Typically found in engines with specialized designs, scraper rings serve the purpose of removing any excess oil from the cylinder walls. This ensures that the engine is only lubricated with the correct amount of oil, helping to maintain efficiency.

Signs of Worn or Damaged Piston Rings

Piston rings can wear down over time due to regular engine operation. Here are some common indicators that your piston rings may need replacement:

- **Reduced Engine Power:** If the engine is struggling to produce power, it may indicate that the piston rings are no longer providing an effective seal, leading to a loss of compression and reduced performance.
- **Increased Oil Consumption:** Worn piston rings allow oil to enter the combustion chamber, resulting in higher oil consumption and potentially causing the engine to burn oil.
- **Excessive Exhaust Smoke:** Blue or gray smoke from the exhaust is a common sign that oil is leaking into the combustion chamber, a direct result of worn piston rings that fail to maintain proper sealing.
- **Poor Fuel Efficiency:** When piston rings fail to maintain the correct compression, the engine has to exert more effort to run, causing a reduction in fuel efficiency as the engine works harder than usual.

Methods of Preserving Piston Rings for Maximum Efficiency

Regular Engine Oil Changes

Keeping your engine oil clean is essential for maintaining the health of the piston rings. Oil that's thick or dirty can cause friction and accelerate wear on the rings.

Routine Engine Inspections

Regularly checking your engine, including the piston rings, for signs of wear or damage can help catch potential issues early and save you from costly repairs down the line.

Avoid Overloading the Engine

Excessive strain on the engine can lead to faster wear and tear on critical components, including the piston rings. Ensure that your tractor is used within its designed limits.

Select the Appropriate Engine Oil

Chossing the suitable oil for your tractor is fundamental. Ensure the oil is reasonable for the environment and kind of work your farm vehicle is doing. Utilizing top notch oils can decrease grating and broaden the existence of cylinder rings.

FLY WHEEL

This mechanical component features a large disc designed to store rotational energy. It provides steady power to the crankshaft, ensuring the engine continues to operate smoothly. The flywheel acts as a temporary energy reservoir, primarily aimed at stabilizing speed variations within the transmission system. By utilizing the stored rotational energy, it helps to smooth the power output from the tractor engine and preserves the crankshaft's angular momentum during the engine's non-power strokes.

Operation of a Flywheel

The operation of the flywheel relies on the principle of inertia. As the engine turns, the flywheel captures energy during the power stroke, when the pistons exert pressure on the crankshaft. This stored energy is subsequently released during the other strokes, assisting in maintaining a steady rotational speed. To visualize this, imagine a seesaw. When one side descends, it accumulates potential energy, which is then released when the opposite side rises. Similarly, the flywheel stores rotational kinetic energy, which is later released to ensure smoother engine operation.

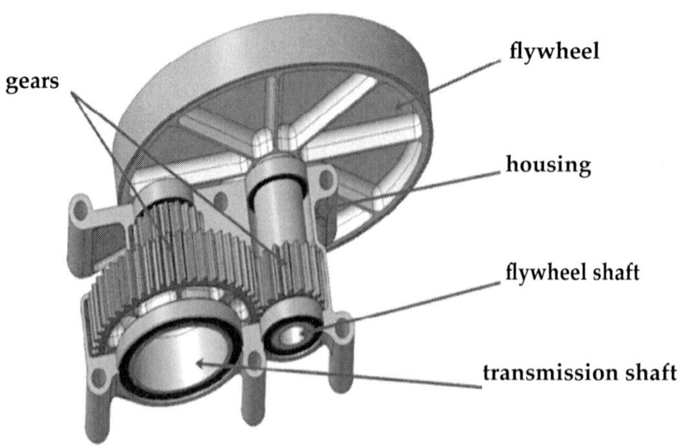

Figure: 2.10 Operation of a Flywheel

Functions of a Flywheel

The flywheel serves several important functions in an engine, including:

- **Energy Storage:** The flywheel functions as a mechanical device that stores rotational energy, accumulating power during times of excess energy production or when the engine is operating efficiently.
- **Clutch Functionality:** Integral to the clutch system, the flywheel helps in the smooth engagement and disengagement of power between the engine and transmission, facilitating seamless gear shifts and vehicle operation.
- **Engine Balancing:** The flywheel plays a key role in balancing the engine and reducing vibrations by counteracting the reciprocating forces created by the pistons and other moving components.
- **Power Regulation:** In certain applications, such as renewable energy systems, the flywheel can regulate power by absorbing excess energy during peak production and releasing it during high-demand periods.
- **Enhancing Crankshaft Rotation:** The flywheel aids in maintaining the rotational momentum of the crankshaft during the engine's non-power strokes, providing stability and preventing potential stalling.
- **Starter Motor Support:** In internal combustion engines, the flywheel is equipped with a ring gear that engages with the starter motor, allowing for smooth engine starts.
- **Smooth Power Delivery:** By capturing energy, the flywheel ensures the consistent delivery of power to the transmission system, contributing to the smooth and uninterrupted operation of the engine.
- **Minimising Speed Fluctuations:** The flywheel's inertia helps reduce fluctuations in the rotational speed of the transmission, ensuring a stable and consistent power output.

Importance of the Flywheel in Engines

The flywheel is essential for maintaining smooth engine operation and effective power delivery. Here's why:

- **Smoother Torque Delivery:** By storing and releasing energy, the flywheel helps reduce torque fluctuations during the engine's cycle, ensuring that power is delivered consistently and smoothly to the transmission.
- **Reduced Vibration:** The flywheel's inertia works to dampen vibrations, allowing the engine to operate more smoothly and quietly.
- **Improved Fuel Efficiency:** By helping to maintain a steady engine speed, the flywheel promotes optimal fuel combustion, which contributes to better fuel efficiency.
- **Engine Starting:** In certain engines, the flywheel provides the necessary initial rotational force to start the engine, facilitating a smooth startup.

Flywheel Size and Torque Delivery

The flywheel's size has a direct effect on its capacity to store and release energy. A larger flywheel can accumulate more energy, leading to smoother torque delivery, particularly at lower engine speeds. This is why engines intended for heavy-duty use, such as those in trucks and buses, typically incorporate larger flywheels.

However, larger flywheels come with certain disadvantages. They tend to increase the engine's inertia, making it slower to accelerate. As a result, smaller flywheels are commonly used in performance-oriented engines, where rapid acceleration is a priority.

FIRING INTERVAL

The firing interval denotes to the time between the ignition of fuel in each cylinder during the engine's cycle. This term is often associated with multi-cylinder engines, where the engine fires multiple cylinders in a set sequence.

Key Details of Firing Interval in Tractor Engines:

1. **Cylinder Ignition Sequence:** In a multi-cylinder tractor engine, fuel is ignited in each cylinder to produce power. The firing interval indicates the time gap between the ignition of fuel in one cylinder and the next, as the engine goes through its firing cycle.
2. **Engine Design and Firing Order:** The firing interval is determined by the firing order—the sequence in which the cylinders of the engine fire. For example, in a 4-cylinder engine, the firing order might be 1-3-4-2, meaning that cylinder 1 fires first, followed by cylinder 3, then 4, and finally cylinder 2. The interval between each firing is dependent on the engine's rotational speed (RPM) and the number of cylinders.
3. **Engine RPM and Firing Interval:** The faster the engine is running (higher RPM), the shorter the firing interval will be, as the engine is completing more cycles per second. The engine's power output and efficiency can be affected by the firing interval, as it plays a role in how smoothly the engine operates.
4. **Impact on Performance:** A well-designed firing interval ensures smooth engine operation, reducing vibration and improving power delivery. If the interval is irregular or incorrect, it could lead to engine misfires, uneven power generation, and rough running.

In short, the firing interval in a tractor engine is the time between the ignition events in different cylinders, and it plays an important role in the engine's smooth operation and efficiency.

FIRING ORDER

Common firing orders are listed below. For V engines and flat engines, the numbering system is L1 for the front cylinder of the left bank, R1 for the front cylinder of the right bank, etc.

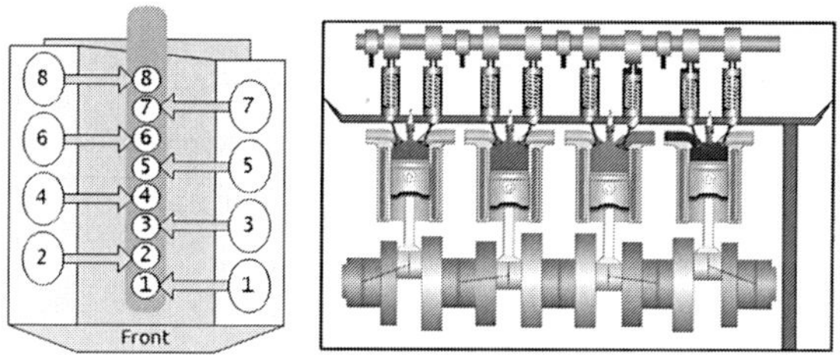

Figure: 2.11 Enging Firing Order

- In two-cylinder engines, the cylinders can fire either simultaneously (as seen in flat-twin engines) or sequentially (as in straight-twin engines).
- In straight-three engines, firing orders such as 1-2-3 and 1-3-2 are essentially equivalent.
- Straight-four engines commonly follow a firing order of 1-3-4-2, although some British engines use 1-2-4-3.
- Flat-four engines usually have a firing order of R1-R2-L1-L2.
- Straight-five engines generally use the firing order 1-2-4-5-3, which helps reduce primary vibration caused by the rocking couple.
- Straight-six engines typically employ a firing order of 1-5-3-6-2-4, resulting in a perfect balance of primary and secondary vibrations.
- V6 engines with a 90-degree angle between the cylinder banks commonly use firing orders like R1-L2-R2-L3-L1-R3 or R1-L3-R3-L2-R2-L1. Meanwhile, V6 engines with a 60-degree angle often use a firing order of R1-L1-R2-L2-R3-L3.
- Flat-six engines generally follow firing orders like R1-L2-R3-L1-R2-L3 or R1-L3-R2-L1-R3-L2.
- V8 engines have various firing orders, with some engines from the same manufacturer using different sequences.
- V10 engines may follow firing orders like R1-L5-R5-L2-R2-L3-R3-L4-R4-L1 or R1-L1-R5-L5-R2-L2-R3-R3-R4-R4.
- V12 engines also utilize a variety of firing orders.

CAMSHAFT

The camshaft is a rotating shaft that controls the intake and exhaust valves during the engine's operating cycle. It regulates the precise timing and duration of when the valves open and close, ensuring proper air intake, fuel combustion, and exhaust gas expulsion. By synchronizing the movement of the valves, the camshaft plays a crucial role in maintaining the engine's power output, fuel efficiency, and overall performance.

CRANKSHAFT

A crankshaft is a rotating component that converts the reciprocating motion of the piston into rotational motion. The power generated by the pistons is transmitted to the crankshaft through the connecting rod. The crankshaft's rotational movement drives the flywheel, ultimately powering the tractor. Typically made from cast iron or forged steel, the crankshaft is a crucial part of the transmission system, and any design flaw can lead to engine malfunction or failure.

FUEL INJECTOR

A fuel injector is responsible for injecting fuel into the combustion chamber at the end of the compression stroke by atomizing the fuel into fine droplets. During the intake stroke in an internal combustion engine, air is drawn into the cylinder, where it is compressed to very high pressure. The fuel is also pressurized to achieve atomization. At the precise moment of injection, a specific quantity of fuel is delivered into the chamber to ensure optimal combustion. The fuel injector plays a critical role in creating a high-quality fuel-air mixture, which enhances engine performance and reduces emissions.

3

Fuel System

INTRODUCTION

The fuel system in a tractor engine is essential for providing the engine with the right amount of fuel to enable efficient combustion. Its primary function is to guarantee a steady and reliable flow of clean, properly atomized fuel, which is crucial for achieving optimal performance and fuel efficiency. The system consists of various components that collaborate to manage different processes such as storing, filtering, and delivering the fuel to the combustion chamber.

These processes must work in harmony to ensure the engine operates smoothly, maintaining both power output and efficiency. Proper atomization of the fuel ensures that it mixes well with air, allowing for more complete combustion and minimizing the chances of fuel wastage or emission of harmful substances. From fuel storage tanks to injectors, every part of the fuel system is designed to meet the demands of the engine under various operating conditions. A well-functioning fuel system helps maintain engine longevity, reduces fuel consumption, and ensures that the engine can perform at its best across different workloads and environments.

Fuel is a substance consumed by the engine to produce power. The common fuel for Internal Combustion engines are

1. Petrol
2. Power kerosene
3. High speed diesel

Calorific value of fuel

The heat liberated by combustion of a fuel is known as calorific value or heat value of the fuel. It is expressed in kcal/kg of the fuel

Sl. No	Name of fuel	Calorific value, kcal/kg
1	Light Diesel Oil (L.D.O)	10300
2	High speed diesel oil (HSD)	10550
3	Power kerosene	10850
4	Petrol	11100

FUEL SUPPLY SYSTEM OF S.I ENGINE

The fuel supply system of sparkignition engine consists of:

1. Fuel tank
2. Sediment bowl
3. Fuellift pump
4. Carburetor,
5. Fuel pipes

In certain spark ignition engines, the fuel tank is positioned higher than the carburettor, allowing fuel to flow by gravity from the tank to the carburettor. Between the fuel tank and carburettor, there may be one or two filters to clean the fuel before it enters the carburettor. Additionally, a transparent sediment bowl is installed to collect any dirt or debris from the fuel. However, if the fuel tank is situated below the level of the carburettor, a lift pump is used to force fuel from the tank to the carburettor. In this setup, the fuel travels from the tank to the sediment bowl, then through the lift pump, and finally to the carburettor via appropriate piping. Once in the carburettor, the fuel is directed into the engine cylinder through the engine's inlet manifold.

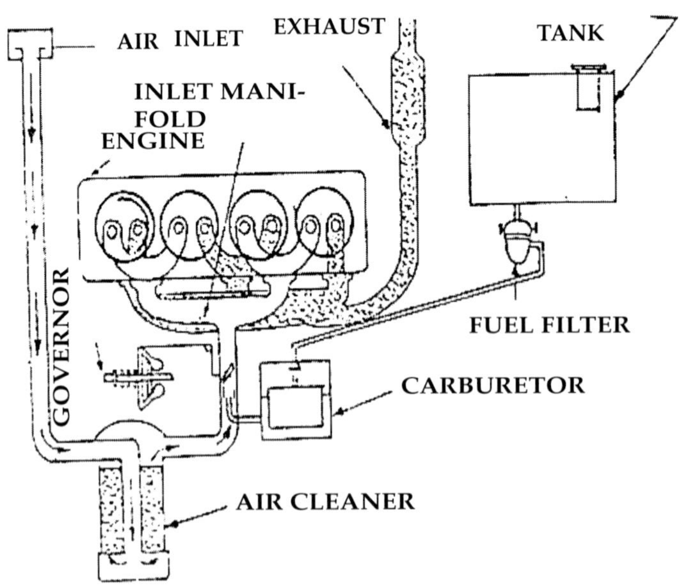

Figure: 3.1 Fuel Supply System for S.I. Engine

Carburettor

The process of preparing air-fuel mixture away from the engine cylinder iscalled carburetion and the device in which this process takes is called carburettor. The main components of a simple carburettor are:

1. Float chamber,
2. Float,
3. Nozzle,
4. Venturi,
5. Throttle valve,
6. Inletvalveand
7. Meteringjet.

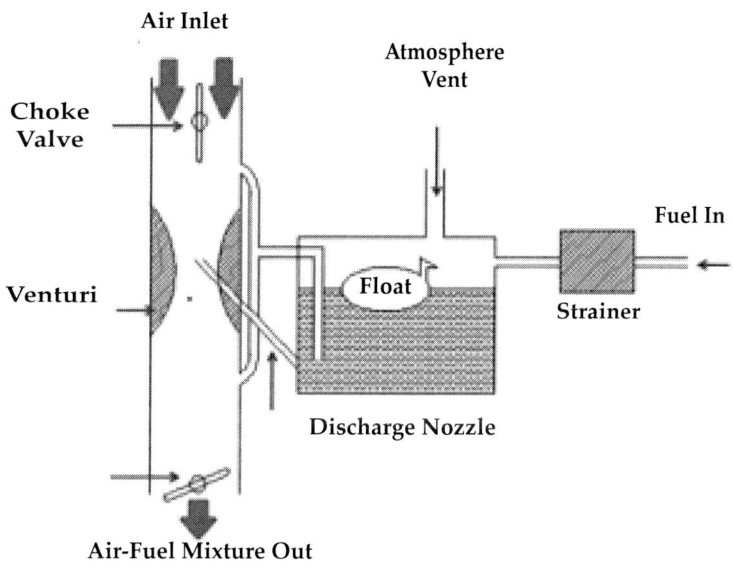

Figure: 3.2 Carburettor

Functions of Carburettor:

1. To thoroughly mix air and fuel
2. To atomize the fuel
3. To regulate the air-fuel ratio at various engine speeds and loads
4. To supply the correct amount of mixture for different engine speeds and loads

FUEL SUPPLY SYSTEM FOR DIESEL ENGINE

The fuel supply system in a diesel engine is comprised of several key components:

1. Fuel Tank
2. Fuel Lift Pump or Fuel Feed Pump

3. Fuel Filter
4. Fuel Injection Pump
5. High-Pressure Pipe
6. Overflow Valve
7. Fuel Injector

- **Fuel Tank:** The fuel tank serves as the storage container for diesel. A wire gauge strainer is located under the cap to prevent the entry of foreign particles into the tank.
- **Fuel Lift Pump:** This pump is responsible for transferring fuel from the fuel tank to the inlet gallery of the fuel injection pump.
- **Preliminary Filter (Sediment Bowl Assembly):** Often installed on the fuel lift pump, this filter is designed to prevent foreign materials from entering the fuel line. It features a glass cap and a gasket.
- **Fuel Filter:** Diesel engines typically utilize a two-stage filtering system:
    - **Primary Filter:** This filter removes coarse materials, water, and dust.
    - **Secondary Filter:** This filter handles the removal of fine dust particles.
- **Fuel Injection Pump:** The fuel injection pump is a high-pressure device that delivers fuel to the injectors in the correct order of engine firing. It generates pressure ranging from 120 kg/cm^2 to 300 kg/cm^2 and ensures that the appropriate amount of fuel is supplied to each cylinder at the right time.
- **Air Venting of Fuel System:** When air enters the fuel lines or the suction chamber of the injection pump, proper venting is required. The priming pump removes air through the bleeding holes in the injection pump.
- **Fuel Injector:** The fuel injector is the component that delivers finely atomized fuel at high pressure to the engine's combustion chamber. Modern tractor engines use injectors with multiple holes. The main parts of the injector include the nozzle body and the needle valve. The needle valve is pressed against a conical seat in the nozzle body by a spring. The injection pressure is adjusted by a screw. During operation, fuel from the injection pump flows into the nozzle body through a high-pressure pipe. When the fuel pressure exceeds the spring pressure, the needle valve lifts off its seat, allowing fuel to be forced out through the nozzle spray holes into the combustion chamber.

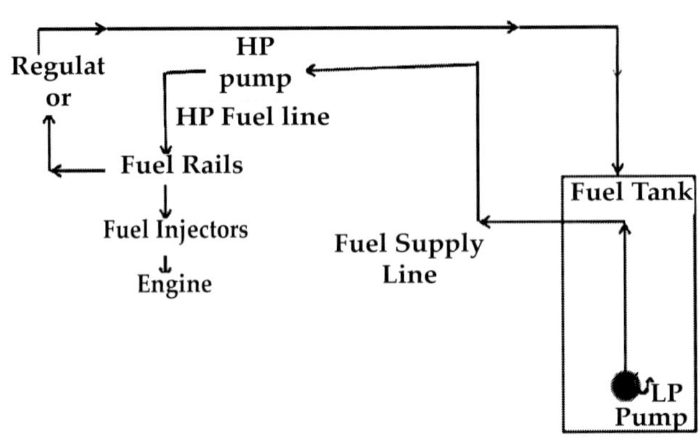

Figure: 3.3 A Fuel Supply System for Diesel Engine

Fuel is drawn from the fuel tank by the fuel feed pump and directed to the injection pump via a fuel filter. The injection pump then delivers high-pressure fuel to the injection nozzles through delivery valves and high-pressure pipes. The fuel is injected into the combustion chamber through these nozzles. Any fuel that leaks from the nozzles is directed through a leakage pipe, where it returns to the fuel tank via the overflow pipe. An overflow valve, located at the top of the filter, regulates the feed pressure to ensure it remains within the specified limit. If the feed pressure exceeds this limit, the overflow valve opens, allowing the excess fuel to flow back to the fuel tank through the overflow pipe.

4

Fuel Pumps, Fuel Filters and Injectors

INTRODUCTION

The fuel system in a tractor engine is essential for maintaining efficient engine operation by ensuring clean and well-regulated fuel is delivered to the combustion chamber for optimal performance. Proper upkeep of this system, including regular checks on fuel filters, lines, and pumps, helps extend the engine's lifespan and guarantees smooth operation during demanding agricultural tasks. The fuel system plays a key role in supplying the engine with precisely regulated fuel for combustion. Core components like the fuel pump, fuel filter, and fuel injectors work together to deliver the right quantity of fuel at the appropriate time, ensuring the engine operates efficiently, performs reliably, and lasts longer.

FUEL PUMP

The fuel pump is responsible for drawing fuel from the fuel tank and delivering it to the engine at the correct pressure. It ensures a consistent and reliable flow of fuel to the fuel injectors or carburetor.

Types of Fuel Pumps

- **Mechanical Fuel Pump:** Common in older tractors, this pump is usually driven by the engine through a gear or camshaft. It delivers fuel from the tank to the fuel system at a steady pressure.
- **Electric Fuel Pump:** Found in modern tractors, electric pumps are powered by the tractor's electrical system and provide more consistent pressure and flow. These are often located near the fuel tank and can be activated by a relay or switch.

Operation of the Fuel Pump

- The fuel pump works by generating a vacuum that draws fuel from the tank into the pump.
- In mechanical pumps, as the engine operates, the diaphragm or piston in the pump moves, drawing fuel into a chamber and then pushing it into the fuel lines at a controlled pressure.

- In electric pumps, the motor drives fuel through a check valve and into the fuel lines, ensuring consistent fuel flow to the injectors.
- The fuel pump guarantees that fuel is supplied to the engine, even under high load or challenging conditions, while maintaining the appropriate fuel pressure for combustion.

Fuel Filter

The fuel filter is essential for ensuring that the fuel supplied to the engine is clean and free from contaminants, which could damage the engine's internal components, such as the fuel injectors and piston chambers.

Working of the Fuel Filter

As the fuel travels from the tank to the engine, it first passes through the fuel filter.

- The filter traps dirt, rust, debris, water, and other impurities in the fuel, preventing them from reaching sensitive engine components.
- Fuel filters use a fine mesh or paper element to catch these contaminants. Some filters are also equipped with a water-separating element to remove any moisture from the fuel before it reaches the injectors.
- In diesel engines, the fuel filter needs to be replaced periodically to ensure that it continues to effectively clean the fuel.

Importance of the Fuel Filter

- Prevents clogging of fuel injectors.
- Ensures that the engine receives clean fuel for efficient combustion.
- Reduces wear and tear on engine components.

Fuel tank

- **It is a storagetank for diesel:** A wire gauge strainer is provided under the cap to prevent foreign particles entering the tank.
- **Fuel lift pump:** It transfers fuel from fuel tank to in let gallery of fuel injection pump.

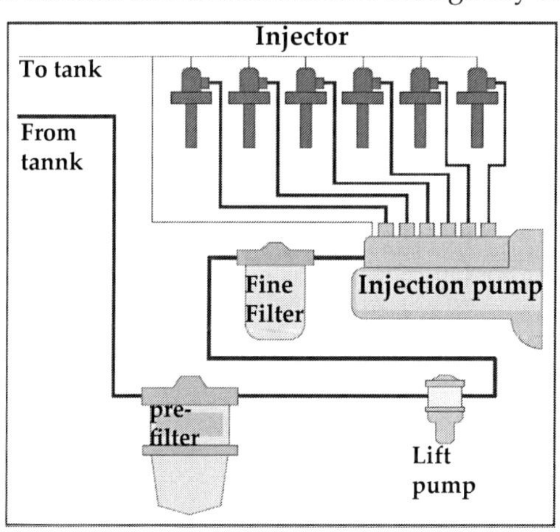

Figure: 4.1 Fuel Supply System Fordiesel Engine

Preliminary Filter (Sediment Bowl Assembly)

This filter is typically installed on the fuel lift pump to prevent foreign materials from entering the fuel line. It consists of a glass cap with a gasket, which serves to trap any debris before the fuel moves further into the system.

Fuel Filter

Diesel engines commonly use a two-stage filtration system:

- **Primary Filter:** This filter removes larger contaminants, such as coarse materials, water, and dust.
- **Secondary Filter:** This filter is designed to remove finer dust particles, ensuring that the fuel entering the engine is clean and free from impurities.

Fuel Injection Pump

It is a high-pressure pump which supplies fuel to the injectors accordingto the firing order of the engine. It is used to create pressure varying from 120 kg/cm^2 to 300 kg/cm^2.Itsuppliestherequiredquantityof fueltoeachcylinderatappropriatetime.

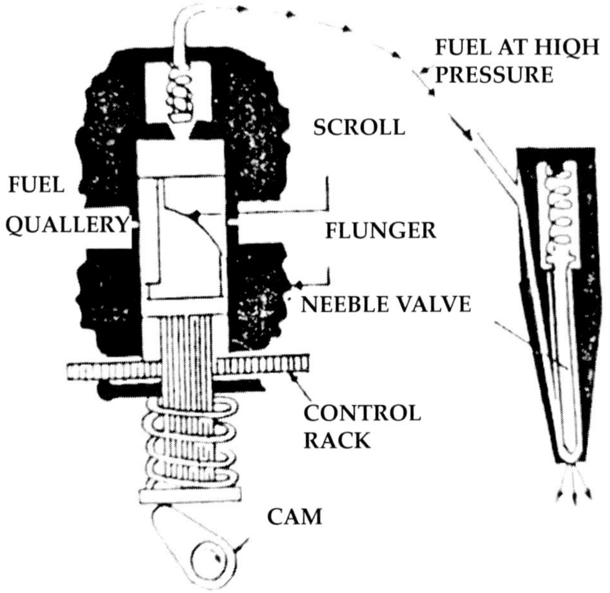

Figure: 4.2 Fuel Injection Pump and Fuel Injector

- **Air venting of fuel system:** When air has entered the fuel lines or suction chamber of the injection pump, venting should be done properly. Airisremovedbytheprimingpump through the bleeding holes of the injection pump.
- **Fuel injector:** The fuel injector is a crucial component responsible for delivering finely atomized fuel under high pressure to the engine's combustion chamber. Modern tractor engines are equipped with fuel injectors that feature multiple

holes for better fuel atomization. The key components of the injector include the nozzle body and the needle valve. The needle valve is pressed against a conical seat within the nozzle body by a spring. The injection pressure is regulated by adjusting a screw. During operation, fuel from the injection pump flows into the nozzle body via a high-pressure pipe. When the fuel pressure surpasses the set spring pressure, the needle valve lifts off its seat, allowing the fuel to be forced out of the nozzle's spray holes into the combustion chamber.

WORKING OF FUEL INJECTION SYSTEM

The functioning of the Fuel Injection system can be described as follows:

- The Engine Control Unit (ECU) plays a crucial role in overseeing the fuel injection process, utilizing data from internal sensors to monitor engine conditions and make necessary adjustments.
- Fuel is sourced from the vehicle's fuel tank and is directed through specialized fuel lines to the injection system.
- Fuel pumps are responsible for pressurizing the fuel, ensuring it reaches the correct pressure for effective injection.
- A fuel pressure regulator is used to ensure the fuel pressure stays within an optimal range for proper engine performance.
- In many systems, a fuel rail is used to distribute fuel evenly to the engine's cylinders.
- The ECU takes full control over the injectors, determining the exact timing and volume of fuel to be injected into each cylinder.
- The fuel-to-air ratio is continuously adjusted based on the engine type, the fuel being used, and the current operating conditions, aiming for improved power output, enhanced fuel efficiency, or stricter emission compliance.

COMPONENTS OF FUEL INJECTION SYSTEM

A fuel injection system consists of essential components such as injectors, fuel pumps, a fuel rail, sensors, and an electronic control unit (ECU). These elements work together to deliver and meter fuel accurately for combustion in the engine.

The Fuel Injection System consists of the following key components:

- **Governor:** The governor regulates the fuel supply to the engine, adjusting it based on the engine's load. This helps maintain smooth and consistent engine performance.
- **Fuel Injector:** Fuel injectors deliver the pressurized fuel from the injection pump into the cylinders, where it combines with air for combustion.
- **Fuel Filter:** The fuel filter removes contaminants such as dirt, debris, and other harmful particles from the fuel, preventing blockages in the injection system and ensuring that clean fuel reaches the engine.
- **Fuel Tank:** The fuel tank acts as the storage reservoir, holding the fuel (gasoline or diesel) until it is required for the engine's operation.

- **Fuel Pump:** The primary role of the fuel pump is to transfer fuel from the tank to the fuel injection system, ensuring a continuous and adequate supply of fuel to the engine.
- **Injection Pump:** The injection pump is responsible for metering and pressurizing the fuel before it is injected into the engine's cylinders, ensuring precise control for optimal combustion.

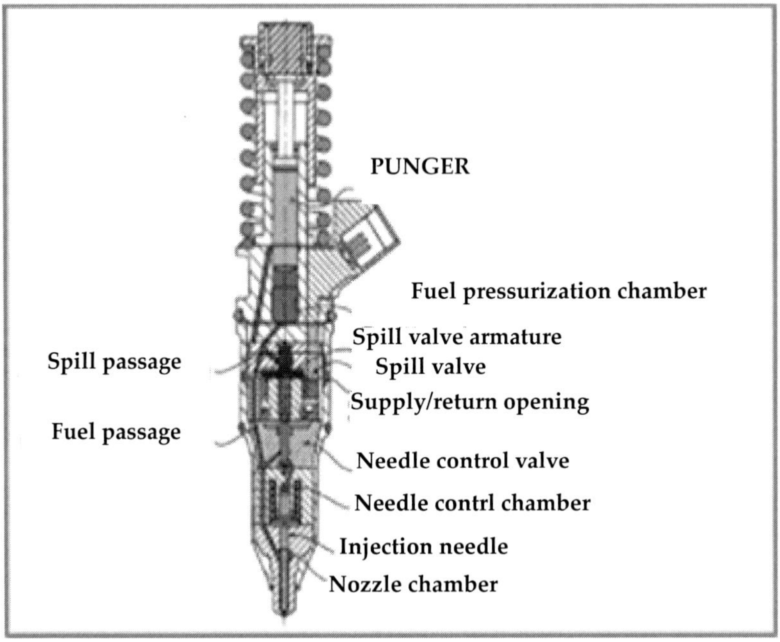

Figure: 4.3 Fuel Injection System

Sensors for Precise Fuel System Control

The various sensors used for precise control of the fuel system in the fuel injection system include:

- **Oxygen Sensor:** This sensor measures the oxygen content in the exhaust gases, allowing the Engine Control Unit (ECU) to assess whether the fuel mixture is rich or lean. Based on this data, the ECU adjusts the fuel mixture to optimize combustion efficiency.
- **Throttle Position Sensor:** Continuously tracking the position of the throttle valve, this sensor provides data on the amount of air entering the engine. The ECU uses this information to adjust the fuel delivery rate, ensuring the correct air-fuel mixture is maintained.
- **Mass Airflow Sensor:** The mass airflow sensor measures the volume of incoming air, helping the ECU determine the exact amount of fuel needed for combustion, ensuring a balanced mixture.

- **Coolant Temperature Sensor:** By monitoring the engine's temperature, this sensor provides the ECU with important information, allowing the ECU to adjust fuel delivery and engine performance based on the engine's operating conditions.
- **Manifold Absolute Pressure Sensor:** Measuring the air pressure in the intake manifold, this sensor provides vital data that the ECU uses to optimize engine performance, ensuring efficient fuel delivery.
- **Engine Speed Sensor:** Monitoring the engine's RPM, this sensor helps calculate the pulse width for fuel injection, ensuring the timing and quantity of fuel injected are accurate for optimal engine performance.
- **Voltage Sensor:** Assessing the system's voltage, the voltage sensor helps the ECU make necessary adjustments, such as increasing idle speed when voltage drops due to increased electrical load, thus maintaining stable engine operation.

TYPES OF FUEL INJECTION SYSTEMS

Automobile fuel injection systems come in a variety of types, each designed to enhance engine performance and efficiency. Systems like Port Fuel Injection and Direct Fuel Injection offer high precision, improving fuel efficiency and reducing emissions. Throttle Body Fuel Injection simplifies the injection process, while Sequential Fuel Injection ensures precise timing for fuel delivery. Common Rail systems and Mechanical Fuel Injection are also used in specific applications to meet diverse engine requirements.

Single-Point or Throttle Body Fuel Injection

Single-Point or Throttle Body Fuel Injection (TBFI) is an early fuel injection system. It combines fuel with incoming air in the intake manifold before distributing it to the cylinders. While TBFI is cost-effective and low-maintenance, it lacks precision. The system can lead to performance issues and decreased fuel efficiency, especially when releasing the accelerator, as excess fuel can be injected.

Multiport Fuel Injection

Multiport Fuel Injection (MPFI) improves upon TBFI by positioning smaller and more accurate injectors closer to each cylinder. This allows for more precise fuel distribution, enhancing fuel efficiency. However, since all injectors spray fuel simultaneously, this can result in unburned fuel during rapid acceleration or deceleration, which might affect overall engine performance.

Sequential Fuel Injection

Sequential Fuel Injection builds on the concept of multiport injection but delivers fuel in precise timing, corresponding to the actions of each cylinder. This method improves both fuel economy and engine performance. As fuel remains in the intake port for a shorter duration, sequential injectors tend to remain cleaner and last longer than other types.

Direct Fuel Injection

Direct Fuel Injection (DFI) involves injecting fuel directly into the combustion chamber, bypassing the intake system entirely. While common in diesel engines, DFI is increasingly used in high-performance gasoline engines. This system offers superior performance and efficiency but requires careful maintenance to ensure optimal engine operation and prevent issues like carbon buildup.

Functions of Fuel Injectors

Fuel injectors perform the following functions:

- Deliver the correct amount of fuel to the engine for optimal combustion.
- Spray fuel into the combustion chamber.
- Are electronically controlled to ensure precise fuel management.
- Ensure the air/fuel mixture is ideal for combustion.
- Enhance engine performance and fuel efficiency.

ADVANTAGES OF FUEL INJECTION SYSTEM

Fuel injection systems in both CI and SI engines offer several benefits:

- Improved fuel efficiency.
- Precise control over the air-fuel mixture.
- Better engine performance.
- Reduced emissions.
- Better cold starting capabilities.
- Lower maintenance needs compared to carburetors.

DISADVANTAGES OF FUEL INJECTION SYSTEM

Fuel injection systems in CI and SI engines have some limitations, including:

- Higher initial cost.
- Complex systems that require skilled technicians for repairs.
- Sensitivity to poor fuel quality.
- Susceptibility to electronic or sensor failures.
- Limited compatibility with older engines.
- Difficulty in diagnosing and repairing issues for some.

APPLICATIONS OF FUEL INJECTION SYSTEM

Fuel injection systems are used in various applications, such as:

- In most modern gasoline and diesel-powered vehicles.
- Commonly found in aircraft engines for precise fuel control.
- Utilized in industrial machinery and generators to improve efficiency.
- Employed in marine engines for better fuel management.
- Applied in certain motorcycles and off-road vehicles to enhance performance.

5

Lubrication System

INTRODUCTION

The lubrication system in an engine plays a critical role in maintaining the smooth operation of its components. It serves to create a barrier between two surfaces that are in contact and moving against each other, reducing the friction that would otherwise lead to wear and tear. To understand the importance of lubrication, consider the effect of friction when two surfaces rub together; much like when rubbing your hands, friction generates heat, which over time can cause damage, such as burns to the skin. In the context of an internal combustion (IC) engine, friction between its moving parts can result in similar consequences, including excessive wear, heat buildup, and power loss, which ultimately impair engine efficiency and longevity.

The internal combustion engine consists of various moving components that are subject to constant friction as they operate. The pistons, crankshaft, camshaft, and valve mechanisms all move at high speeds within confined spaces, leading to the gradual wearing away of metal surfaces. Without proper lubrication, this friction would not only damage the engine's components but also result in overheating and a significant reduction in the engine's performance. The heat generated by this friction can increase the risk of engine failure, as it can lead to parts seizing or warping under high temperatures.

To combat these issues, the lubrication system is designed to reduce friction, dissipate heat, and protect the moving parts from wear. It ensures that a thin film of lubricant is constantly supplied to critical areas of the engine, forming a protective layer between metal surfaces. This helps to minimize direct contact between the moving components, thereby reducing friction and preventing unnecessary wear. Moreover, lubrication contributes to cooling the engine by carrying heat away from the friction zones and dispersing it throughout the engine system.

The effective functioning of the lubrication system is essential for maintaining the overall health and performance of the engine. It not only extends the lifespan of engine parts but also ensures smoother operation, enhanced power output, and greater fuel efficiency. Without a proper lubrication system, the engine would be prone to rapid deterioration, significantly compromising its reliability and performance over time.

Purpose of Lubrication

1. Reducing frictional effect
2. Cooling effect
3. Sealing effect
4. Cleaning effect

LUBRICANTS

Lubricants can be derived from animal fats, plants, and minerals. Vegetable-based lubricants are sourced from seeds, fruits, and plants, with examples including cottonseed oil, olive oil, linseed oil, and castor oil. Mineral lubricants, which are the most commonly used for engines and machinery, are obtained from crude petroleum. These petroleum-based lubricants are more affordable and well-suited for internal combustion engines.

LUBRICATION SYSTEM

The lubricating system of an engine is an arrangement of mechanisms which maintains the supply of lubricating oil to the rubbing surfaces of an engine at correct pressure and temperature.

The parts which require lubrication are:

1. Cylinder walls and piston]
2. Piston pin
3. crankshaft and connecting rod bearings
4. Camshaft bearings
5. Valve operating mechanism
6. Cooling fan
7. Water pump and
8. Ignition mechanism

Ignition System

The ignition system in an internal combustion engine is responsible for igniting the fuel mixture in the engine cylinder at the appropriate time to enable effective operation. The arrangement of components that ensures ignition at the correct timing is referred to as the ignition system.

Types of ignition systems include:

1. Electric spark ignition (spark ignition)
2. Compression ignition (ignition by heat of compression)

3. Hot tube or hot bulb ignition
4. Open fire ignition

Among these, only the first two methods are commonly used in modern engines.

Spark Ignition

The purpose of spark ignition is to provide a precisely timed burst of electricity across an open gap in each cylinder, igniting the fuel-air mixture at the optimal moment. This ensures that the combustion process begins with maximum efficiency, contributing to the engine's overall performance. There are two primary types of spark ignition systems: a) Battery ignition, and b) Magneto ignition. Both systems serve the same fundamental function but differ in their mechanisms of generating and delivering the spark.

Magneto Ignition System

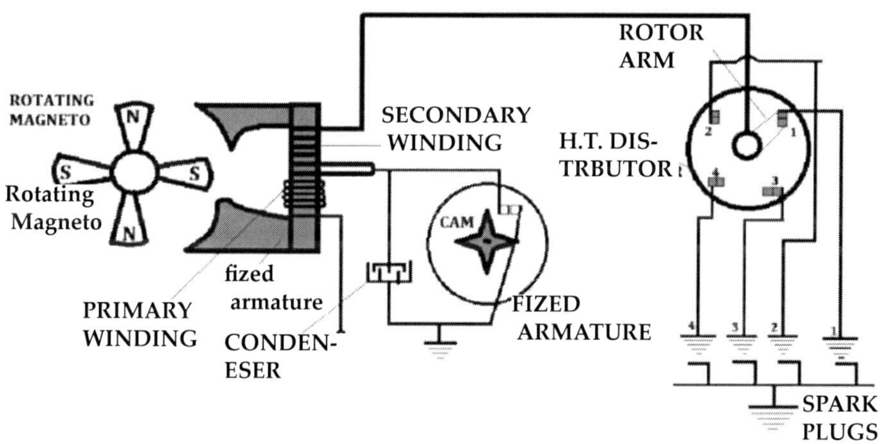

Figure: 5.1 Magneto Ignition System

In magneto ignition system a magneto is used to generate electric current for producing spark. A high tension magneto generates a very high voltage needed for spark plug

Main Components of Magneto Ignition System

- Frame
- Permanent magnet
- Armature
- Soft iron field
- rotor

- Primary and secondary winding
- Breaker points
- Condenser

The armature consists of an iron core with two sets of windings: primary and secondary. Driven by the engine, the armature rotates, causing the primary windings to cut through the magnetic field lines, which induces a current in the primary circuit. As the primary current reaches its peak in both directions, the contact breaker opens the primary circuit abruptly, causing the current to collapse. This sudden collapse induces a high voltage in the secondary winding, generating a brief spark across the spark plug gap. A distributor is used to direct the current to the spark plugs via high-tension wires. The condenser serves to prevent arcing at the breaker points and helps intensify the current in the secondary circuit. For multi-cylinder engines, a distributor and rotor are necessary to distribute the current to the appropriate spark plugs.

ENGINE LUBRICATION SYSTEM

The engine lubrication system is a vital mechanism that ensures a constant supply of lubricating oil to the engine's moving parts at the correct pressure and temperature. Internal combustion engines consist of numerous moving components, and due to the continuous interaction of metal surfaces, friction, wear, heat generation, and power loss occur. Lubrication is essential to prevent these adverse effects and ensure smooth engine operation.

Lubricants are derived from various sources, including animal fats, vegetables, and minerals. Vegetable lubricants are obtained from plant-based sources like cotton seed oil, olive oil, linseed oil, and castor oil. Animal fats, while usable, are not suitable for high-temperature conditions. Mineral lubricants, derived from crude petroleum, are the most common and affordable option for engine applications, especially in internal combustion engines.

The engine lubrication system plays several critical roles:

1. It minimizes power loss by reducing friction between moving parts.
2. It reduces the wear and tear of these parts, prolonging engine life.
3. It provides cooling effects to prevent overheating of hot engine components.
4. It cushions vibrations caused by engine operation.
5. It aids in internal cleaning by removing contaminants.
6. It helps the piston rings create a seal against high-pressure gases in the cylinders.

The lubrication system supplies oil to various parts of the engine, including the crankshaft main bearings, big-end bearings, piston pins and small-end bushes, cylinder walls, piston rings, timing gears, camshaft and bearings, valves, tappets and push rods, oil pump parts, water pump bearings, fuel injection pump bearings, turbocharger bearings (if present), vacuum pump bearings (if present), and air compressor pistons and bearings (in commercial vehicles with air brakes).

TYPES OF LUBRICATING SYSTEMS

There are two main types of lubrication systems used in automotive engines:

1. Splash System
2. Forced Feed System (Pressure System)

Splash Lubrication System

In a splash lubrication system, oil is drawn from the oil sump or pan by a dipper attached to the bottom of the connecting rod for the piston in each cylinder. The connecting rod picks up the oil from the sump and moves it upwards to lubricate various engine components. Additionally, some of the oil reaches different parts like the bearings and connecting rods through the mechanical turbulence generated in the crankcase by the moving components. This turbulence causes the oil to spread as a mist throughout the crankcase. The mist then reaches other components such as the piston, piston pin, and cylinder walls. The camshaft and valve mechanism are also lubricated by this oil.

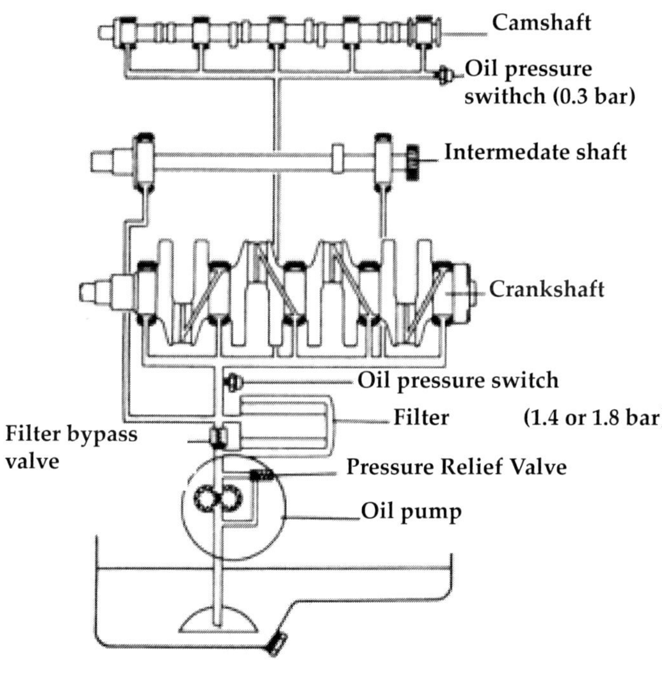

Figure: 5.2 Splash Lubrication System

Although the splash lubrication system is still used in some engines, its application is now mostly limited to smaller or single-cylinder engines. This is because splash lubrication alone does not provide sufficient oil distribution to all the critical components of the engine. As a result, it has been largely replaced by more advanced systems in larger, more complex engines.

Force Feed Lubrication System

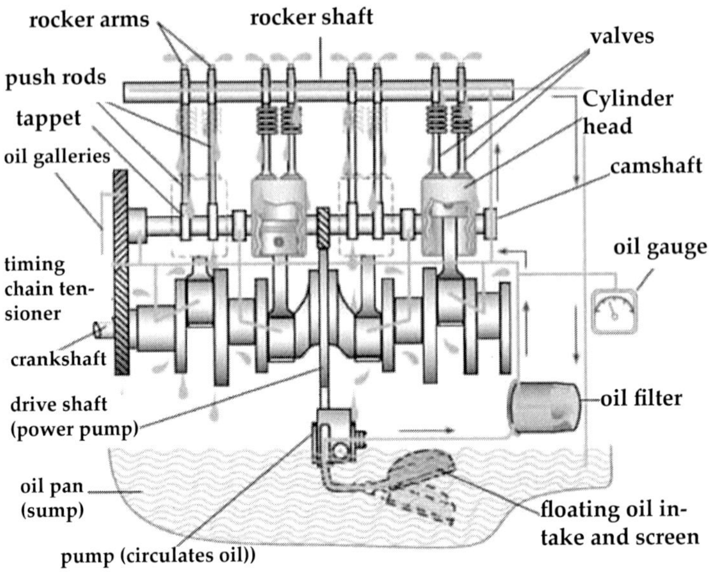

Figure: 5.3 Forced Feed System of Engine Lubrication

Due to the limitations of the splash system in adequately lubricating all critical engine components, the force-feed system is employed to generate additional pressure, ensuring oil reaches all essential parts for lubrication. Typically, a gear-type pump driven by the camshaft is used to create the necessary pressure. This pump moves the oil from the crankcase to the crankshaft, connecting rods, bearings, pistons, and valves. By supplying lubricating oil under pressure, the system enhances the oil's ability to reach even the most distant engine components. As a result, this leads to more efficient lubrication, ultimately contributing to improved engine performance and longevity.

OILPAN/SUMP

Figure: 5.4 Oil Soaked Air Cleaner

An oil pan, or sump, is a bowl-shaped reservoir designed to store engine oil and circulate it throughout the engine. Positioned beneath the crankcase, the oil sump holds the engine oil when the engine is off. During operation, oil is returned to the sump via pressure or gravity. The oil pan's location at the bottom of the engine allows it to collect and store oil efficiently. However, rough road conditions can potentially damage the oil pan. To protect it, manufacturers often install a stone or sump guard underneath, which absorbs impacts from uneven surfaces, safeguarding the sump from harm.

Air Cleaner

- **Definition:** It is a device which filters and removesdust, moisture and other foreign matter from the air before it reaches the engine cylinder.Air cleaners are of three type:
  1. Oil soaked element type
  2. Dry type air cleaners and
  3. Oil bath type air cleaner

Oil Soaked Element Type

The oil-soaked type air cleaner consists of a container filled with wire screens that are saturated with lubricating oil from the crankcase. When the air stream passes through, dust particles are trapped by the oil film. This type of cleaner is commonly used in smaller sizes for crankcase ventilation systems and on stationary engines that operate in relatively low-dust environments. The filtering element requires periodic cleaning, typically by blowing off the accumulated dirt with compressed air. If excessive dirt has built up, the element should be washed with kerosene and then cleaned with pressurized air to remove all remnants of kerosene. Afterward, it should be thoroughly soaked in fresh oil before being reinstalled.

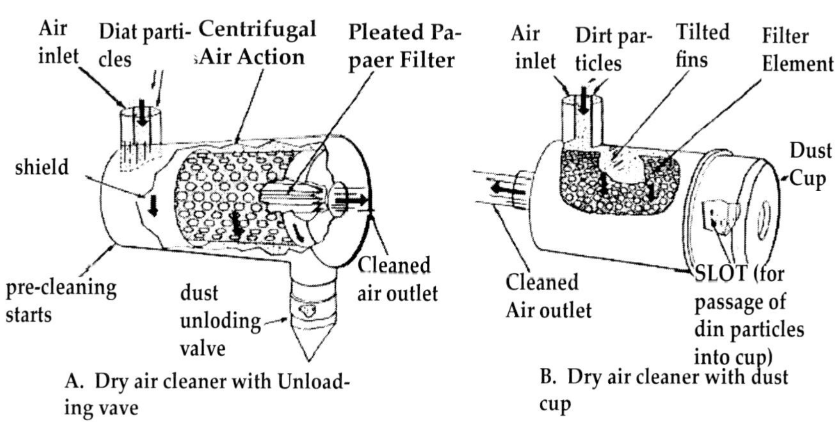

Figure: 5.5 Oil Bath Type Air Cleaner

Dry Type Air Cleaner

These types of air cleaners consist of three main components: the pre-cleaner, the main housing, and the cleaning element, all sealed together as one unit. The filtering element in this system is a dry type of filter. As air passes through the element, the larger surface area of the filter reduces the airspeed. This reduction in speed causes particles or dirt in the air to be deposited on or trapped by the surface of the filter.

Oil bath Type Air Cleaner

In this type of air cleaner, the incoming air impinges upon surface of oil, kept in a container in the lower part of casing. The foreign particles of the air are trapped in the oil and then air passes through a wire element before reaching the inlet manifoldof the engine. The wire element also arrests the remaining dirt particles of the air.

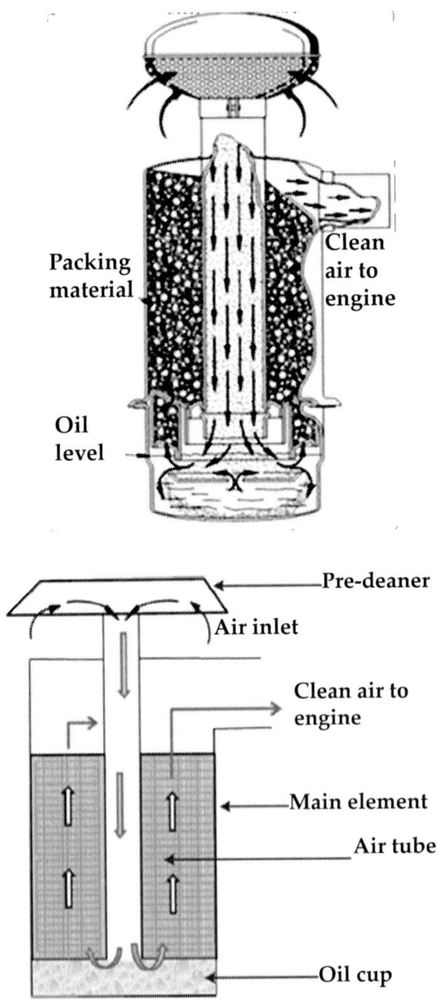

Figure: 5.6 Oil Bath Type Air Cleaner

6

Oil Filters and Oil Pumps

INTRODUCTION

A tractor engine oil filter operates by directing engine oil through a pleated paper filter element that captures dirt, metal debris, and other contaminants. This process ensures that only clean oil returns to the engine, preventing internal parts from experiencing excessive wear and tear by maintaining a clean lubrication system. The filter is typically equipped with a bypass valve to maintain oil circulation even if the filter becomes partially clogged, thus safeguarding the engine from potential damage due to oil starvation. Essentially, the engine oil filter is designed to continuously purify the oil, removing harmful impurities such as particles created by the constant friction between moving parts, dust from environmental contamination, and carbon residues from combustion.

Figure: 6.1 Oil Filter

OIL FILTER

An oil filter is a mechanical device designed to remove contaminants, dirt, and unburned fuel from the circulating oil in an engine. Its primary function is to ensure that only clean and fresh oil enters the engine while trapping and retaining dirt and impurities. Regular maintenance or replacement of the oil filter is essential to maintain the efficiency and longevity of the engine.

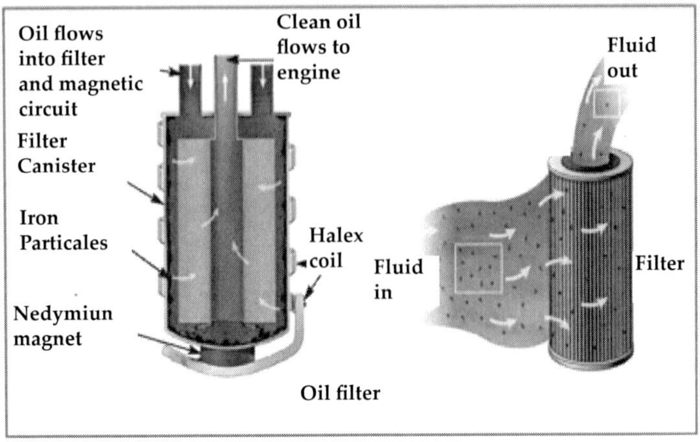

Figure: 6.2 Structure and Function of Oil Fliter

In the early stages of engine development, oil filters were either nonexistent or of poor quality, which led to the necessity of frequent oil changes. This problem was eventually addressed with the creation of the first full-flow oil filtration system. In this conventional setup, oil passes through the filter before reaching the engine's moving components, ensuring cleaner oil for the engine's operation. Oil filters are not only vital in automotive engines but are also used in other hydraulic systems, such as automatic transmissions and power steering. Additionally, gas turbine engines, including those in jet aircraft, rely on oil filters to ensure efficient filtration. From cars, trucks, motorcycles, boats, airplanes to tractors, a wide range of vehicles and machinery depend on oil filters to purify the oil, ensuring smooth engine performance by removing contaminants before the oil circulates through the system.

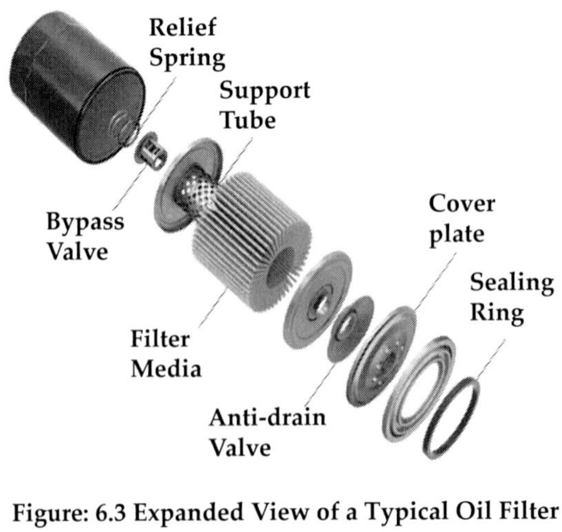

Figure: 6.3 Expanded View of a Typical Oil Filter

The oil filter and oil must work in tandem to protect the engine from harmful microscopic metal flakes produced by combustion, along with dirt and dust. Installing a substandard filter or leaving a filter in place for too long can be as damaging as running an engine with old, ineffective oil. A filter that does not perform efficiently can undermine the protective benefits of high-quality oil, accelerating engine wear, reducing reliability, and shortening the engine's lifespan. The key takeaway is that the oil's capacity to protect the engine's moving parts from wear is directly linked to the filter's effectiveness in removing contaminants from the oil. The filter capture efficiency quantifies the effectiveness of the filter in eliminating particles from oil. A filter must possess sufficient holding ability to retain all dirt until the subsequent oil and filter change. Vehicles functioning in dusty conditions may benefit from filters with significantly greater dirt-holding capacity, or operators may need to replace filters more frequently.

To comprehend how a filter eliminates and keeps particles, it is essential to grasp the filter's design and the dynamics of oil movement in and out of the canister.

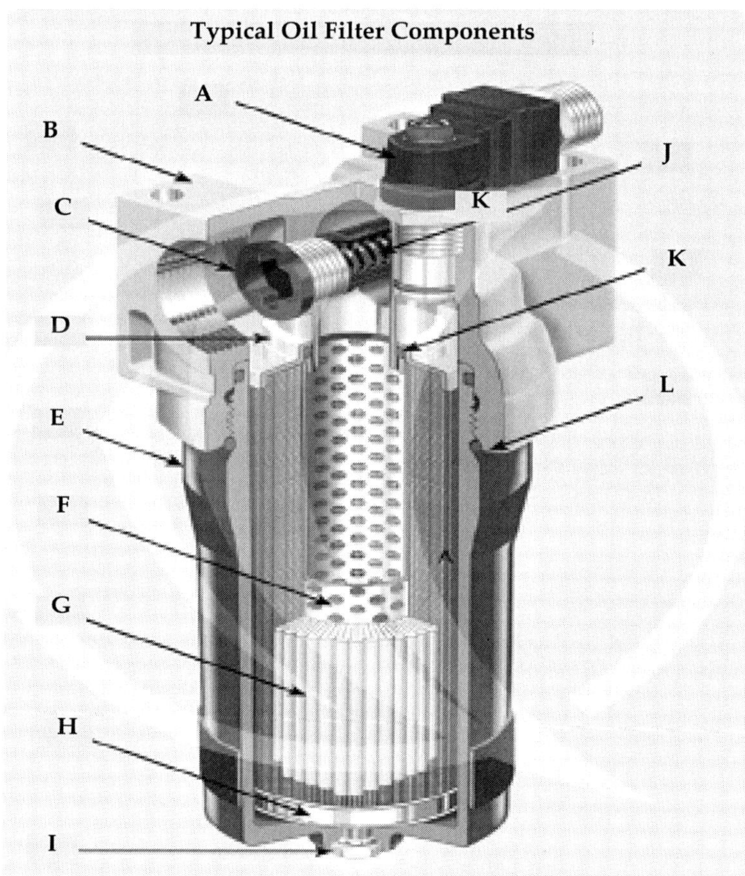

Figure: 6.4 Typical Oil Filter Components.

A. **Condition Indicator:** This device typically measures the differential pressure in order to indicate the remaining life or failure of the oil filter.

B. **Filter Head:** The top enclosure for the filter case contains ports for inlet and outlet flow as well as bypass and pressure differential indicators.

C. **Bypass Valve:** Although there are many type and position variations for bypass valves, the valve shown is tensioned by a spring to allow partial or full fluid flow when the differential pressure exceeds a predetermined pressure.

D. **Baseplate:** This is the filter media's supporting structure on the interfacing end of the filter head. It helps prevent leaks or ruptures due to increased differential pressures and often contains threads for connecting to the filter head.

E. **Filter Housing (Bowl):** The main enclosure for the filter helps direct the oil flow through the filter element.

F. **Center Tube (Inner Core):** This is the central duct for outlet flow from the filter media to the filter head. The center tube also acts as a support for the filter media to prevent collapse from increased differential pressure.

G. **Filter Media:** A pleated filter element is an example of a surface-type filter, which provides a large surface area of porous cellulose or synthetic material for efficient filtration during normal flow conditions.

H. **End Cap:** This is the filter media's supporting structure on the opposite end of the filter head. It helps prevent leaks or ruptures due to increased differential pressures.

I. **Drain Port:** This port provides the option to have the oil drained prior to removal of the oil filter from the filter head. It can also act as a location to take an oil sample or remove excess oil prior to disposal.

J. **Coil Spring:** This is the predetermined tension for the bypass valve. In other configurations, a leaf spring may be used.

K. **Baseplate Gasket:** This pre-lubed sealing ring or molding O-ring provides sufficient sealing between the baseplate and filter head during the filter installation.

L. **Dust Seal:** This prevents dust and other contaminants from entering the interior of the filter housing.

How an Oil Filter blocks

The oil filter is positioned between the oil pump and the engine. When the engine operates, it draws oil from the oil pan through a wire cloth strainer, and then pumps it to the filter. Once inside the filter, the oil passes under pressure through a smaller inlet port that forms a concentric circle around the filter's opening at the top. Most oil filters are designed with an outside-to-inside flow, meaning that oil enters from the outside of the pleated filter elements and flows inward toward the center tube. The filter medium, made from materials like paper, glass fibers, or a blend of both, captures impurities as oil passes through.

Inside the filter, the oil follows the path of least resistance. It first travels through the larger pores of the filter media and into the perforated center tube. Over time, as the larger pores become blocked with contaminants, the oil is forced through smaller pores, and eventually, all pores will become clogged. The filtered oil exits through the center port and moves through a series of manifolds to various parts of the engine. After lubrication, the oil flows back into the pan by gravity, starting the cycle anew.

To prevent damage to the engine when the filter becomes clogged, a mechanical feature is integrated into the system to redirect the oil flow. A fail-safe bypass valve on the engine block ensures that oil continues to flow even if the filter is severely obstructed. However, during this bypass mode, the oil circulating through the engine will not be filtered, making it crucial to replace the filter as soon as possible to prevent engine damage.

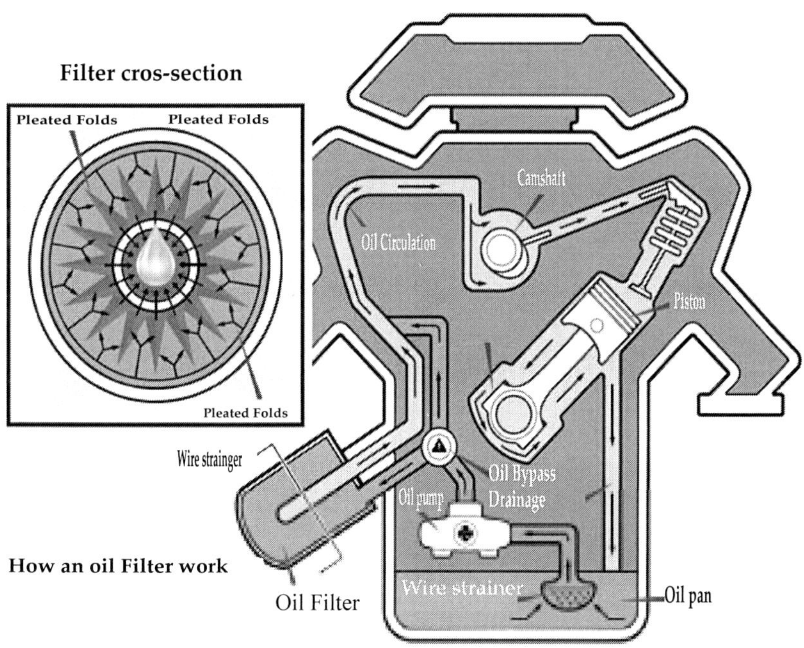

Figure: 6.5 Working of An Oil Filter

PRIMARY FUNCTIONS OF OIL FILTERS

Tractor oil filters play a crucial role in maintaining engine cleanliness by filtering out impurities such as dirt, sediment, and metal particles from the engine oil. This filtration process ensures that the engine's internal components remain protected from wear and damage. The primary functions of tractor oil filters include:

- **Preventing Contamination:** Oil filters trap harmful contaminants in the oil, helping to keep the engine free from debris.
- **Reducing Wear:** By ensuring the oil is clean, the filter prevents metal parts from rubbing against each other, reducing friction and wear.
- **Enhancing Engine Performance:** With filtered oil, the engine operates more efficiently, leading to better performance.
- **Extending Oil Life:** Clean oil lasts longer, which reduces the frequency of oil changes, helping to extend the life of both the oil and the engine.

TYPES OF OIL FILTERS

Oil filters are essential for preserving engine health by removing harmful contaminants from the lubricating oil. There are various types of oil filters, each serving a distinct purpose. Spin-on filters, commonly found in cars, are self-contained units that are easy to replace. Cartridge filters, on the other hand, are replaceable filter elements housed within a larger casing. Magnetic filters work by attracting metal particles, helping to capture metallic debris. Centrifugal filters utilize centrifugal force to separate contaminants from the oil. Each type is designed to effectively keep engines clean and running smoothly by ensuring that only purified oil circulates through the engine.

Full-Flow Oil Filter

The full-flow oil filter, often referred to as the primary oil filter, is the most widely used type in automobiles. It effectively cleans and removes contaminants from all the oil circulating through the engine, making it especially suitable for colder climates where oil may thicken during cold starts. Its full-flow design ensures that oil moves freely through the engine, providing consistent lubrication to maintain optimal engine performance and prevent wear and tear. This type of filter plays a crucial role in maintaining engine efficiency and longevity.

Cartridge Oil Filter

Figure: 6.6 Cartridge Oil Filter

The cartridge oil filter is a streamlined variant of the full-flow filter, positioned vertically for convenient examination without necessitating oil extraction. It is offered in both metal and fiber varieties, with fiber filters being more prevalent due to their recyclability.

Spin-On Oil Filter

Figure: 6.7 Spin-On Oil Filter

The spin-on oil filter comprises a steel canister with a paper element. Its straightforward installation and little tool necessity render it user-friendly, allowing automobile owners to replace it with ease.

Secondary or By-Pass Oil Filter

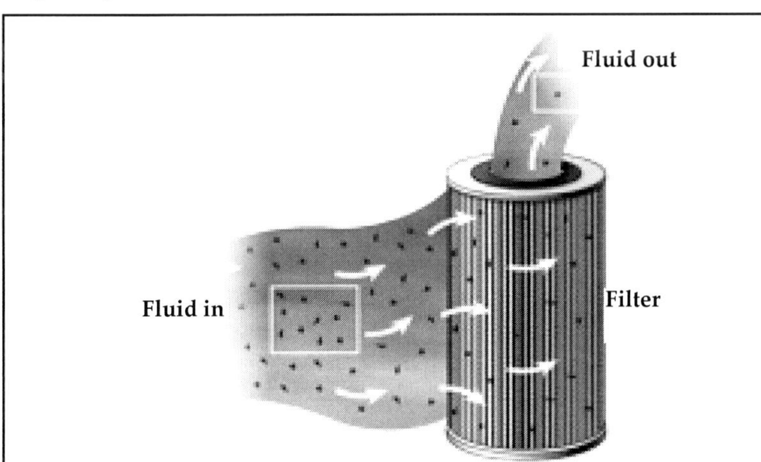

Figure: 6.8 Secondary or By-Pass Oil Filter

Secondary oil filters are utilized in conjunction with full-flow filters to offer further assistance. They refine a minor portion of the engine oil, eliminating impurities that the full-flow filter could have overlooked. These filters prolong the lifespan of motor oil and provide additional safety for the engine.

Spinner Oil Filter

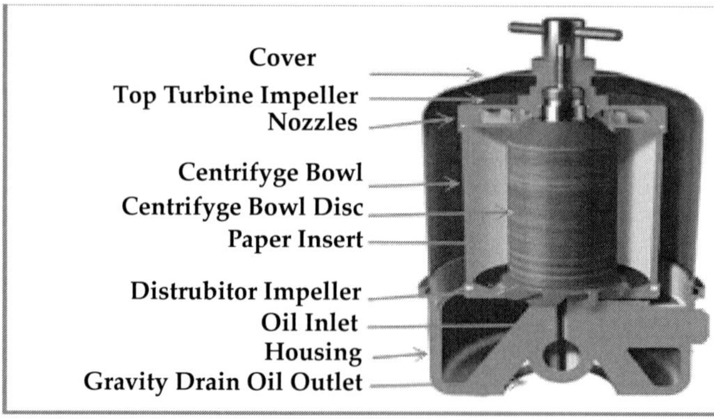

Figure: 6.9 Spinner Oil Filter

A spinner oil filter is a secondary filtration device that utilizes centrifugal force to capture impurities from the engine oil. Certain variations can produce forces above 2,000 times that of gravity, efficiently eliminating even the minutest impurities from the motor oil.

Magnetic Oil Filter

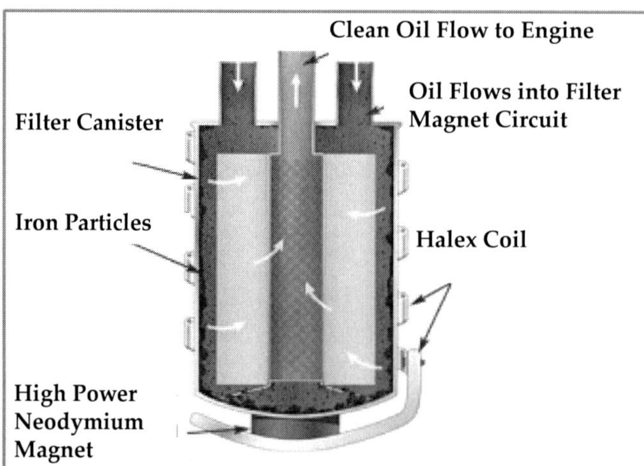

Figure: 6.10 Magnetic Oil Filter

The magnetic oil filter, akin to the spinner filter, serves as a supplementary filter utilized with the full-flow variant. It effectively eliminates metallic impurities from the motor oil but fails to remove dust and dirt. Magnetic filters necessitate no replacement and require only routine cleaning to preserve their effectiveness.

Selecting the right oil filter is essential for maintaining the engine's health and performance in your tractor. Using the wrong filter can result in engine damage and serious operational issues.

Important Factors for Choosing the Right Oil Filter:

- **Tractor Model and Engine Type:** Choose an oil filter that matches the specific model and engine type of your tractor. Always consult the manufacturer's manual for the recommended filter.
- **Filter Material:** The quality of the filter material is key to ensuring efficient filtration. Filters made from high-quality materials tend to be more durable and offer superior performance.
- **Original Parts:** It's best to use original oil filters whenever possible. These filters are designed for optimal compatibility with your tractor's engine and often come with a manufacturer's warranty.
- **Aftermarket Products:** Aftermarket oil filters can be a viable option, but it's important to select filters from reliable brands and ensure the products meet quality standards.

APPLICATIONS OF OIL FILTERS

Oil filters play a vital role across multiple industries and machines, helping to preserve the quality and efficiency of lubricating oils. In automotive applications, they ensure that engines operate smoothly, while in industrial settings, they protect hydraulic systems by filtering out contaminants. Specialized oil filters are also essential for aviation and marine engines, safeguarding critical components and ensuring safe and efficient performance.

- In automotive engines, both spin-on and cartridge oil filters are commonly employed in cars, trucks, motorcycles, and other vehicles. These filters effectively remove impurities, helping protect engine components from wear and tear.
- For industrial machinery, oil filters are crucial in equipment such as generators, construction machinery, and agricultural tools. They maintain the quality of lubricants, contributing to the longevity of these machines and preventing costly repairs.
- In the marine industry, oil filters are indispensable in boats and ships to prevent corrosion and engine damage. These filters are key to ensuring reliable and safe engine operation on the water, where the conditions can be harsh.
- Aircraft engines also rely on oil filters to keep engine oil clean, which is crucial for maintaining the engine's performance and ensuring the safety of flight operations.

Finally, in power generation, oil filters are used in generators and power plants to protect engine and generator performance by preventing damage caused by contaminants in the lubricating oil. This helps to maintain efficiency and reliability in power production.

WORKING PRINCIPLE OF OIL FILTER

The operation of oil filters is simple. Oil enters through small openings along the outer base flange and passes through the filter element to be cleaned. The filtered oil then exits through a larger central hole and returns to the engine.

Most modern oil filters now feature an anti-drain-back valve, typically made of rubber and located around the base flange's perimeter holes. This valve opens to allow oil into the filter under pressure and closes when the engine is off, preventing oil from draining out. This mechanism ensures that the oil remains in the filter, preventing dry starts, where the engine starts without adequate lubrication.

Many modern internal combustion engines also incorporate a secondary oil filter, known as a bypass filter, for added protection. This filter allows oil to bypass the main filter under certain conditions, such as extremely cold weather when oil may become too thick. It also permits unfiltered oil to pass through if the main filter becomes clogged, thus preventing engine oil starvation and ensuring continuous lubrication even under challenging circumstances.

MAINTENANCE AND REPLACEMENT OF OIL FILTERS

Regular maintenance and timely replacement of tractor oil filters are essential for ensuring the long-term, efficient operation of the engine. Failure to maintain the oil filter can result in reduced engine performance and higher chances of malfunction.

- **Change Intervals:** Adhere to the oil filter change intervals outlined in your tractor's manual. Typically, it is recommended to replace the oil filter every 200-300 operating hours.
- **Oil Change Along with Filter Change:** It is advisable to replace the oil filter during an oil change to ensure both components work at optimal levels.
- **Regular Inspection:** Conduct regular inspections of the oil filter. If the filter shows signs of clogging or damage, replace it immediately.
- **Professional Service:** Have the oil filter replaced by an authorized service provider to ensure it is done correctly and safely.

Tips for Filter Maintaining

Maintaining the oil filter is an easy but effective way to keep your tractor running smoothly. Below are some tips for filter maintenance:

- **Follow the Manufacturer's Recommendations:** Always refer to the tractor's manual for specific filter replacement intervals and specifications. These guidelines are created to ensure peak performance and durability.
- **Inspect Filters Regularly:** Perform visual inspections, particularly before critical farming seasons or when operating in dusty conditions. If you notice any dirt buildup, tears, or other damage, replace the filter right away.
- **Use Quality Replacement Parts:** When replacing filters, select high-quality aftermarket options like those from Vapormatic. These filters meet industry standards and provide reliable performance at a more affordable price than OEM filters.
- **Store Filters Properly:** If keeping spare filters, store them in a clean, dry environment to avoid contamination before installation. Filters stored improperly can accumulate dust and moisture, impairing their effectiveness.

Impact of Oil Filters on Engine Performance

The effect of oil filters on engine performance cannot be overstated. Clean oil promotes smoother engine operation and helps optimize fuel usage.

- **Ensures Efficient Operation:** Clean oil enables efficient engine operation, boosting overall performance.
- **Provides Fuel Savings:** A well-maintained engine uses fuel more efficiently, contributing to fuel savings.
- **Minimizes Wear:** Clean oil helps prevent excessive wear of internal engine components, ensuring the engine lasts longer.
- **Reduces Maintenance Costs:** Consistently changing the oil filter prevents major engine malfunctions, ultimately lowering maintenance costs.

Tractor oil filters play a crucial role in engine performance and durability. Choosing the right filter, keeping it well-maintained, and replacing it when needed ensures the tractor functions efficiently and reduces operating costs. It's essential to recognize the importance of oil filters on engine performance and address their environmental impact by using proper disposal methods. This contributes to sustainable farming practices. By following these steps, you can protect the engine and improve the efficiency of your agricultural activities.

ADVANTAGES OF OIL FILTERS

Oil filters offer several significant benefits, including:

- **Efficient Contaminant Removal:** Oil filters effectively eliminate contaminants and impurities, ensuring cleaner oil circulation throughout the engine.
- **Prolongs Engine Life:** By removing abrasive particles, oil filters prevent engine parts from wearing out and causing damage, thus extending engine life.
- **Improves Engine Performance:** Consistent and adequate lubrication provided by oil filters enhances engine performance.
- **Enhances Fuel Efficiency:** A cleaner engine operates more efficiently, resulting in improved fuel efficiency.
- **Maintains Optimal Oil Pressure:** Oil filters help maintain proper oil pressure, ensuring smooth operation and reducing the risk of engine failure.
- **Reduces Emissions:** By aiding the proper functioning of the engine, oil filters contribute to cleaner exhaust gases and lower emissions.

DISADVANTAGES OF OIL FILTERS

Despite their importance, oil filters have some limitations that should not be ignored:

- **Clogging or Blockage:** Over time, oil filters can become clogged, reducing oil flow and potentially damaging the engine.
- **Inefficient Filtration:** Low-quality or inefficient filters may not remove all contaminants, allowing harmful particles to circulate in the engine.
- **Limited Capacity:** Some filters have a limited filtration capacity and need to be replaced more frequently.

- **Increased Maintenance Costs:** Oil filters can contribute to higher maintenance costs, especially if premium or specialized filters are required.
- **Neglected Maintenance:** Poorly maintained filters can become a source of engine problems, affecting reliability and performance.
- **Defective Filters:** In rare instances, faulty filters can cause oil leaks or other malfunctions in the engine system.
- **Environmental Concerns:** Some oil filters are not environmentally friendly due to the materials used in their construction or improper disposal methods.
- **Compatibility Issues:** Using the wrong type or size of oil filter can cause compatibility problems and damage the engine.
- **Oil Flow Restrictions:** Certain filters may restrict oil flow excessively, causing a drop in oil pressure and negatively impacting engine performance.

DIRT-HOLDING CAPACITY AND CAPTURE EFFICIENCY

While oil flowing through a filter is a standard process, not all filters are equally effective at trapping and holding debris from the oil. The efficiency of an oil filter can differ significantly across manufacturers and even among different brands from the same company (for instance, between budget-friendly filters and high-performance ones). One thing is clear: Seeing terms like "deluxe," "high-performance," "super-efficient," or "top-quality" on a filter doesn't necessarily indicate that you're purchasing a superior product.

Figure: 6.11 Dirt-Holding Capacity of Oil Filter

There are two criteria that influence oil filter performance and cost:

1. Dirt-holding capacity
2. Capture efficiency

Oil Filters and Oil Pumps

The Qualties are Ascertained when the filter is evaluated according to ISO 4548-12—Methods of Test for Full-flow Lubricating Oil Filters for Internal Combustion Engines. This ISO testing procedure techiniques offers essential information that allow us to make side-by-side comparisons of various filters.

Dirt-Holding Capacity

A filter's dirt-holding capacity determines how long it will function before entering bypass mode. Filters designed for synthetic oils and rated for 15,000 miles will typically last longer than those intended for shorter intervals like 3,000 or 5,000 miles. When selecting a filter, consider purchasing one rated for the distance you plan to drive between oil changes, with a margin of safety. Additionally, take into account the dustiness of your environment—dusty conditions require filters that can trap more dirt.

Another method for choosing the right filter is to check its ability to trap debris. The ISO test quantifies this by measuring the grams of dirt the filter can hold before it becomes ineffective. There's a significant difference between filters that capture 14 grams of dirt versus those that capture 28 grams. Filters with higher dirt-holding capacities are typically more expensive.

In the ISO test, technicians gradually introduce test dust upstream of the filter while monitoring oil pressure. As the filter accumulates dirt, the oil pressure increases slightly. When the pressure reaches a terminal drop level set by the manufacturer, the test concludes. The total amount of dirt collected by the filter is calculated from the amount introduced.

Capture Efficiency

The beta ratio is another key indicator of oil filter quality, as it reflects the filter's ability to capture particles of various sizes. Different manufacturers use filters with media that have varying average pore sizes. Finer pores are necessary to capture smaller, micron-sized particles, but these finer pores tend to be more expensive than coarser media, which have larger pores and are less costly.

The test conditions for determining the beta ratio are identical to those used for assessing capture efficiency. Both tests are performed simultaneously. In the test, conducted on a laboratory test stand, oil flows from the test reservoir through the filter and then back into the same reservoir.

During the test, controlled laboratory test dust is introduced while two special sensors, known as particle counters, are placed both upstream and downstream of the filter. As the filter captures particles of different sizes, the concentration of particles upstream of the filter will always be higher than that downstream, reflecting the filter's efficiency in trapping contaminants.

The beta ratio is the ratio of these two concentrations:

$$\text{beta ratio} = \frac{\text{number of particles upstream}}{\text{number of particles downstream}}$$

For any Specify particle size (eg. 10 micrometers) the increased the beta ratio, the better the filter's capture efficiency. For Example, if ten particles greater than 10 microns (micrometers) are counted upstream of the filter and only one of these particles is counted downstream, then the beta ratio is 10 (10/1).

You can convert the Beta Ratio to efficiency as:

$$\text{capture efficiency (\%)} = \frac{\text{beta ration -1} \times 100}{\text{beta ratio}}$$

The beta ratio and the capture efficiency have no value unless we know precisely the size of the microns being evaluated in the test (10 microns for instance). Working film clearance between the piston ring and cylinder wall in engines is extremely small — maybe as little as 10 microns.

Debris in the 10-micron range can cut or scratch metal as the particles roll between opposing surfaces. Particles larger than 10 microns will not fit into the smaller gaps and will be swept aside. Smaller particles will generally flow through the gap with the oil.

Consider that a standard oil filter removes particles that are larger than 40 microns. It's been demonstrated that particles that average 10 microns can produce four times more wear on rods, rings, and bearings than particles larger than 20 microns.

OIL PUMPS

The oil pump, typically a gear-type pump, is responsible for pushing oil into the oil pipe. It is powered by the engine's camshaft. The lower part of the pump extends into the crankcase and is covered with a screen to filter out foreign particles. A portion of the oil is directed to the oil filter, while the remaining oil circulates to lubricate various engine components.

An oil pressure gauge placed in the line measures the pressure within the lubrication system, typically around 3 kg/cm^2 in a tractor engine. If the oil pressure gauge indicates low pressure, it suggests a fault in the system that must be addressed immediately. The lubricating oil pump operates as a positive displacement pump.

Functions of an Oil Pump

Here are some essential functions of an oil pump:

- It circulates oil and ensures that an engine functions smoothly with sufficient lubrication.
- It ensures that engines do not undergo wear and tear due to constant subjection to friction.

- It circulates the oil in a specific direction through engine galleries.
- It stores the hot oil in the reservoir

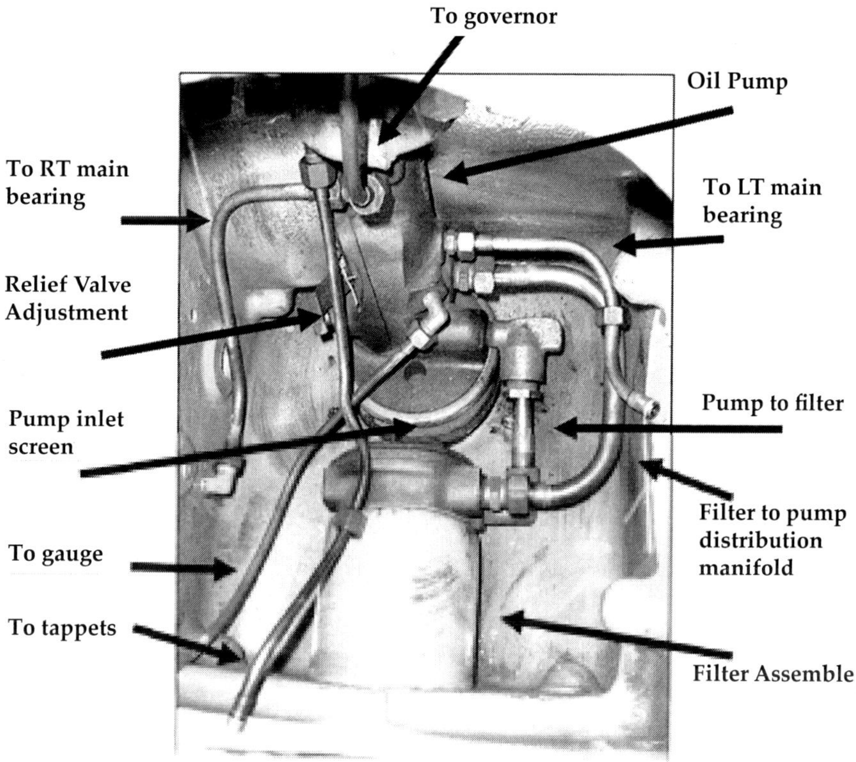

Figure: 6.12 Structure and function of Oil Pump with Pipes

Types of Oil Pumps

The three most popular types of oil pumps commonly available in the market are as follows:

- **Rotor Oil Pump:** Rotor pumps are typically located inside the crankcase, making it easier to draw and circulate oil throughout the engine. They feature inner and outer gears that rotate in unison. For optimal performance, the inner gear has one fewer lobe and is slightly offset. This design requires precise calculation and measurements to maintain effective operation.
- **Twin Gear Oil Pump:** Positioned within the oil pan at the engine's base, the twin-gear oil pump uses two intermeshing gears to pump oil. A shaft connected to the camshaft, crankshaft, or gear moves the first gear, which in turn drives the second gear, ensuring efficient oil circulation to engine components.
- **Front Cover Oil Pump:** Also called internal or external pumps, front cover oil pumps are similar to rotor pumps in structure. They consist of an outer rotor

and an inner gear connected directly to the crankshaft. This eliminates the need for an additional pump driveshaft and allows the pump to rotate at the same speed as the engine, producing high oil pressure.

Working Principle of an Oil Pump

Oil pumps play a critical role in circulating oil throughout an engine once the engine starts running. Here's a breakdown of how oil pumps operate:

- **Step 1:** The oil first passes through a strainer that filters out large particles or debris.
- **Step 2:** Next, the oil enters the pump, which generates pressure to circulate the oil throughout the engine.
- **Step 3:** The oil then flows through a heat exchanger for cooling before proceeding further.
- **Step 4:** The oil passes through galleries, which direct it to key engine components, including the crankshaft, piston, and camshaft.
- **Step 5:** Finally, the oil returns to the sump, a reservoir where it is stored for future use.

In some engines, an injector takes a portion of the oil, sealing the space between the piston ring and cylinder walls to prevent air from escaping. This improves the overall engine efficiency.

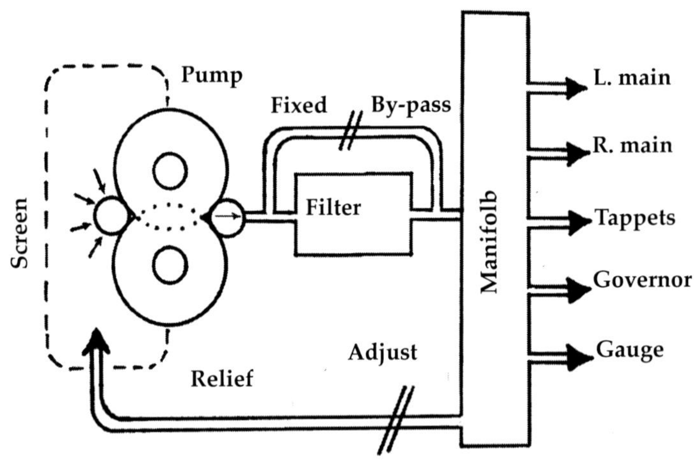

Figure: 6.13 Schematic Diagram of Engine Oiling

Symptoms of Oil Pump Failure

When an oil pump fails, it often results in the inability to circulate oil properly throughout the engine, leading to severe degradation in engine performance. Here are several symptoms that indicate the potential failure of an oil pump:

1. **Flashing Oil Light Indicator:** A flashing oil light typically signals an issue with the oil pump. The pump controls the oil pressure and ensures smooth circulation. A malfunctioning pump can cause oil pressure to drop, triggering the warning light.

2. **Elevated Engine Temperature:** The oil pump plays a crucial role in circulating oil to prevent friction. When the pump fails to supply sufficient oil, the engine experiences increased friction and wear, leading to a rise in both internal and external engine temperatures. This excess heat can cause damage to the engine and its components.
3. **Noisy Valve Train:** The valve train, consisting of parts like hydraulic lifters, valve guides, push rods, and seals, relies on oil for lubrication. When the oil pump fails to provide adequate lubrication, friction increases between these components, producing a noise indicative of an oil pump issue.
4. **Noisy Oil Pump:** A whirring sound from the engine, even when the engine is not running, suggests a severe malfunction in the oil pump. Although rare, this noise is a critical warning and requires immediate action, typically involving the replacement of the oil pump.
5. **Noisy Hydraulic Lifters:** Hydraulic lifters, essential for the proper functioning of the valve train, need proper oil lubrication. When the oil pump fails to deliver sufficient oil, the lifters produce noise due to the lack of lubrication.

In conclusion, the oil pump plays a vital role in engine performance. While technological advancements continue to improve the efficiency and reliability of oil pumps, it's essential to remain vigilant about the pump's condition and address any issues promptly to maintain engine health.

Lubricating Pump Features

- Robust in Nature
- Optimized Design
- High flexibility
- Durable
- Reliable
- Low noice or vibrations
- Low maintenance costs

He provide choice of Oil Pumps campalible with Tractors, Trucks, Diesel Engines etc encompassing both single cylinder and multi-cylinder engine oil pumps.

Pumps are Available in a Variety of Stock and Custom Configurations.

The simplest pump is a single-stage external wet sump engine oil pump, which uses one or two sections to generate the necessary flow and pressure. In two-section pumps, there is a bearing section placed between the two engine oil sections. A turbocharger oil or diesel fuel section can be added to this single-stage pump, with an independent regulator installed to manage the additional section. Alternatively, both turbocharger oil and diesel fuel sections can be incorporated into the engine oil pump. Delivery capabilities range from 12 to 40 gallons per minute at 3000 RPM of the pump shaft, depending on the housing selection and configuration. For custom needs, various components can be combined beyond these standard setups to create a tailored pump.

7

Cooling System

INTRODUCTION

In an internal combustion engine, fuel is combusted within the cylinder to generate power. During the power stroke, the temperature can reach up to 1600 ºC, which exceeds the melting point of engine components. The optimal operating temperature for internal combustion engines ranges between 140 ºF and 200 ºF, making cooling a crucial aspect of engine operation. It is estimated that approximately 40% of the total heat generated is released into the atmosphere through the exhaust, 30% is dissipated through the cooling system, and the remaining 30% is used to produce power.

The cooling system in internal combustion engines plays a vital role in ensuring the engine operates efficiently and reliably. It is responsible for maintaining the engine's temperature within the optimal range, preventing overheating that can lead to engine damage. The engine generates substantial amounts of heat during combustion, and without proper cooling, these temperatures can rise to levels that exceed the melting point of engine components. This can result in engine failure, reduced performance, and significantly shortened engine life. The primary function of the cooling system is to absorb excess heat from the engine and dissipate it into the atmosphere, keeping the engine temperature within safe limits. This allows the engine to perform at its best while avoiding costly repairs and downtime.

The cooling system works by circulating coolant through the engine block, where it absorbs heat, and then transferring this heat to the radiator, where it is cooled by airflow or a fan. There are various types of cooling systems, including air-cooled and liquid-cooled systems, each offering distinct advantages depending on the engine's design and application. Liquid cooling systems, which are more common in modern vehicles, typically offer more effective heat dissipation, making them ideal for high-performance engines. Proper maintenance of the cooling system is crucial for the engine's long-term health, as it ensures that all components are functioning optimally, such as the radiator, thermostat, water pump, and hoses. Neglecting the cooling system can lead to severe consequences, such as overheating, engine misfires, and even complete engine failure. Understanding the cooling system's components and their functions is essential for anyone working with engines, from mechanics to vehicle owners, to ensure optimal engine performance and longevity.

PURPOSE OF COOLING

1. To maintain the optimum temperature of the engine for efficient performance under all conditions.
2. To disperse surplus heat for the protection of engine components such as the cylinder, cylinder head, piston, piston rings, and valves
3. To retain the lubricating properties of oil withen the engine

METHODS OF COOLING

1. Air cooled system
2. Water cooled system

Aircooling System

An air blower or fan is used in tractors equipped with air-cooled engines to circulate air and dissipate heat from the engine surface, especially the cylinder. Special baffles and fins are designed to direct the air towards the heated components, ensuring that no hot spots form. The size and spacing of the fins are influenced by factors such as the amount of heat to be removed, air temperature, air speed, the material of the fins, and the size of the cylinder. Typically, a greater number of short fins is preferred over fewer, larger fins. To ensure effective cooling, the surface area of the metal in contact with the air needs to be maximized. When the fins are closely spaced, higher air pressure is necessary. Materials like copper and steel alloys, known for their excellent heat conductivity, are also used to enhance heat transfer. Air-cooled engines are lighter and quicker to warm up compared to water-cooled engines and are especially beneficial in extreme weather conditions where water may freeze.

In air-cooled engines, heat is directly conducted from the engine components to the atmosphere. The principle of air cooling involves the use of fins on the cylinder to increase the surface area, facilitating faster cooling. The cylinder is typically enclosed in a sheet metal casing known as cowling. The flywheel is fitted with blades on its face, functioning like a fan that draws air through an opening in the cowling, directing it around the finned cylinder. To maintain the air-cooled system, it is crucial to keep the air passages clear by removing debris, such as grass, using a stiff brush or compressed air.

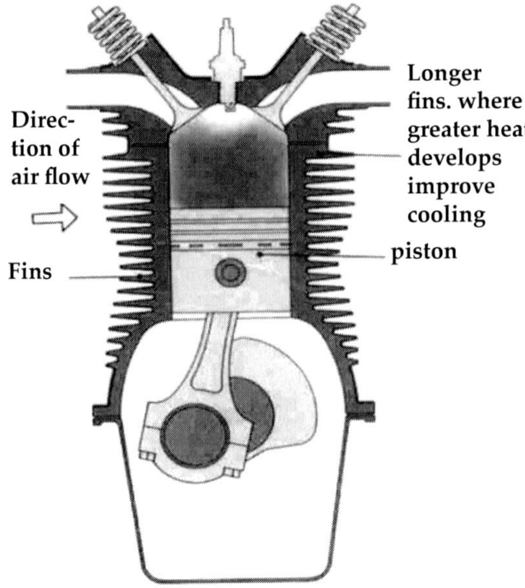

Figure: 7.1 Cooling Fins are used to Increase the Suface Area Exposed to the Airstream

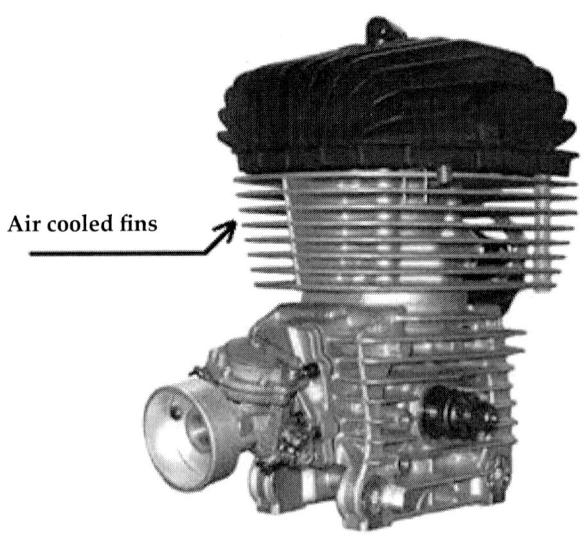

Figure: 7.2 Aircooled Engine

Advantages of Air Cooling System in IC Engine

The advantages of Air Cooling in IC Engines include:

- Lighter engine design.

- Simplicity and lower cost.
- Reduced maintenance needs.
- Suitable for smaller engines like motorcycles and scooters.
- No risk of freezing-related damage.

Disadvantages of Air Cooling System in IC Engine

Some major limitations of Air Cooling Systems are:

- Less precise temperature control.
- Can be noisier due to fans.
- Limited cooling efficiency.
- Not ideal for high-performance or larger engines.
- Higher risk of overheating.

Water Cooling System

In a water cooling system, water jackets are placed around the engine cylinders or liners. The circulating water in these jackets absorbs heat from the cylinder surfaces, and the heated water is then cooled by air passing through the radiator. The system includes components such as water jackets, a water pump, radiator, thermostat valve, fan, belt, and pulley. While water is the most commonly used cooling agent, specialized coolants with enhanced properties like corrosion resistance and higher boiling points are also available and recommended for maintaining better engine efficiency.

The water circulates continuously through the water jackets at the required pressure and speed, facilitated by a water pump driven by a belt. Typically, centrifugal water pumps are used, consisting of an inlet, outlet, and an impeller that forces the water out through the outlet via centrifugal force. The pump's inlet connects to the bottom of the radiator, drawing coolant or water from it. When the engine cools, the thermostat valve remains open, allowing the same water or coolant to circulate through the jackets. As the coolant heats up, the thermostat opens to let the water pass through the radiator, where it dissipates heat by coming into contact with the air flowing through the radiator. The radiator, located at the front of the vehicle, consists of a coolant tank, tubes, and a pressure cap, which prevents water evaporation and increases internal system pressure. The temperature difference between the outside air and the water inside the radiator accelerates the heat transfer.

Engines typically operate efficiently within a temperature range of 80°C to 90°C, and it is desirable for the engine to reach this temperature quickly in cool weather, maintaining it within this range during hot weather. The thermostat controls this temperature range by regulating the coolant circulating through the water jackets. Proper maintenance of the engine cooling system is essential for optimal engine efficiency and fuel economy. Operating the engine at too low a temperature can cause fuel loss due to unburnt fuel and oil dilution, while high temperatures can lead to excessive wear and engine seizure.

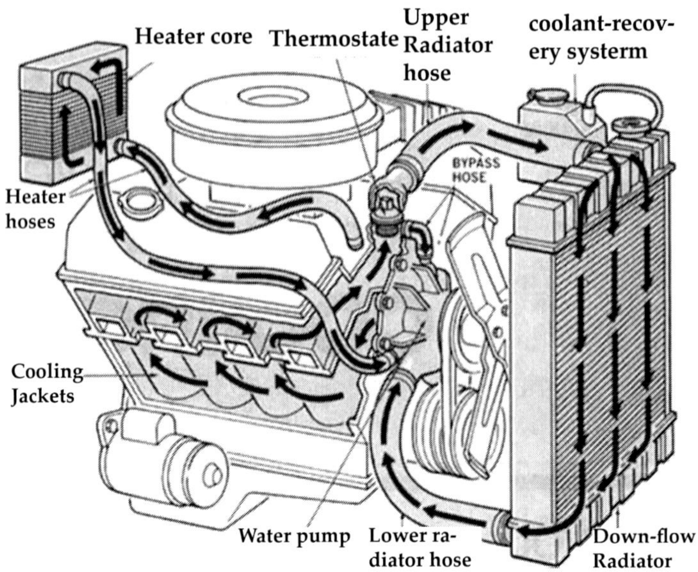

Figure: 7.3 Water Cooled Engines

Engines that utilize water as cooling medium are referred to as water cooled engines. Water is circulated round the cylinders to absorb heat from the cylinder walls. The heated water is conducted through a radiator to remove the heat and cool the water.

Method Sofwater Cooling

1. Openjacketorhoppermethod
2. Thermo-siphon method
3. Forced circulation method

Open Jacket Method

A hopper or jacket encasing water surrounds the engine cylinder. As long as the hopper has water, the engine continues to function well. Once the water reaches a boiling point, it is substituted with cold water.The hopper is enough large to operate for several hours without requiring a refill.The Adrian plug is situated in a conveniently accessible location for water drainage when needed.

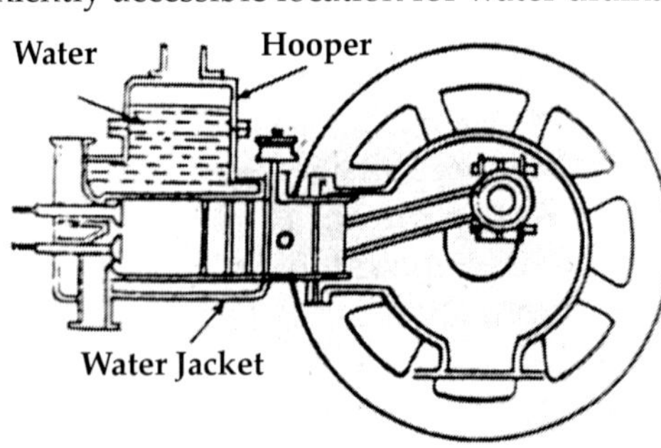

Figure: 7.4 Open Jacket Water Cooling System

Thermo-siph on method

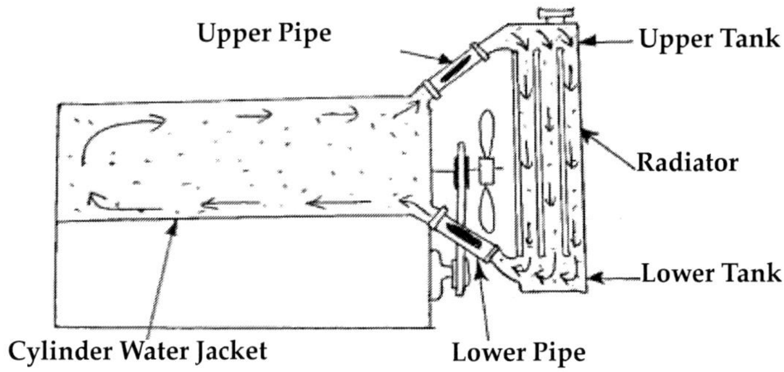

Figure: 7.5 Thermo-siph on Method of Cooling Engines

It comprises a radiator, water jacket, fan, temperature gauge, and hose connections.The method operates on the idea that heated water surrounding the cylinder gets less dense and ascends in a liquid column. Hot water circulates through the radiator, traversing tubes enveloped by air. The circulation of water occurs because the water jacket and radiator are connected at both the top and bottom.

A fan is operated by a V-belt to draw air through the radiator unit's tubes, thereby cooling the radiator water. The system's drawback is that the circulation of water is significantly diminished due to the buildup of scale or foreign substances in the channel, resulting in engine overheating.

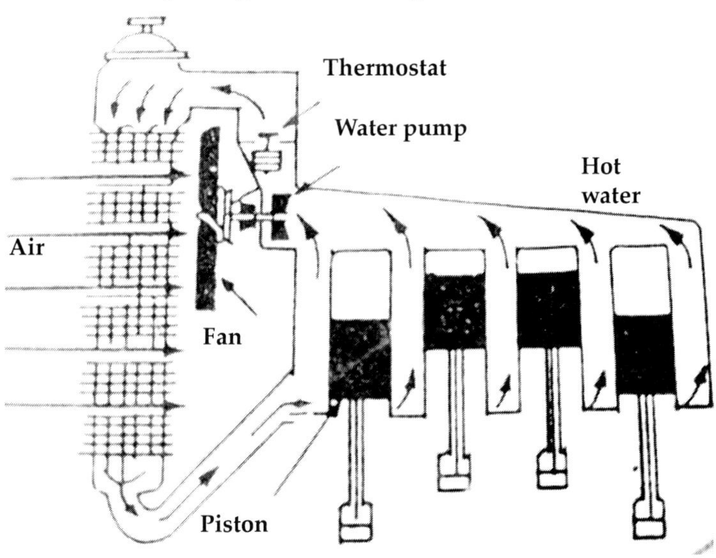

Figure: 7.6 Forced Circulation System

Forced Circulation System

In this system, a water pump is utilized to push water from the radiator into the engine's water jacket. After circulating through the entire water jacket, the water returns to the radiator, where it cools down through radiation. To maintain the optimal engine temperature, a thermostat valve is installed at the outer end of the cylinder head. Initially, the coolant is bypassed through the water jacket until the engine reaches the desired temperature. Once this temperature is reached, the thermostat valve opens, closing the bypass and directing the coolant to flow into the radiator for cooling. This forced circulation system consists of several key components: the water pump, radiator, fan, fan belt, water jacket, thermostat valve, temperature gauge, and hosepipe. The system consists of the following components:

1. Waterpump,
2. Radiator,
3. Cooling
4. Fan,
5. Fan-belt.
6. Water jacket,
7. Thermostat valve,
8. Temperature gauge,
9. Hosepipe.

- **Water Pump:** A centrifugal pump is used to speed up the rate of water circulation in the cooling system. It draws the cooled water from bottom of the radiator and delivers it to the water jackets surrounding the engine.
- **Radiator:** The radiator transfers the heat absorbed by the water to the surrounding air.

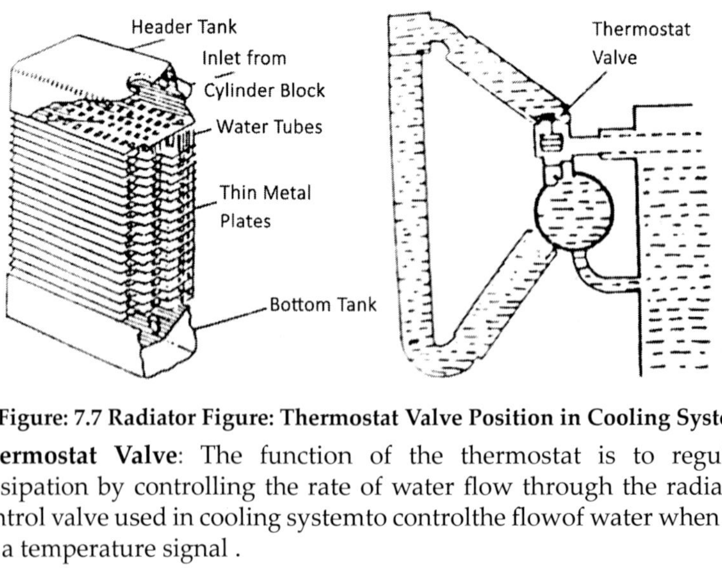

Figure: 7.7 Radiator Figure: Thermostat Valve Position in Cooling System

- **Thermostat Valve**: The function of the thermostat is to regulate heat dissipation by controlling the rate of water flow through the radiator. It isa control valve used in cooling systemto controlthe flowof water when activated by a temperature signal .

- **Fan-belt:** A fan-belt is a forced circulation cooling system of an internal combustion engine drives the fan to circulate cooling, maintaining optimal engine temprature and preventing over heating.
- **Cooling Fan:** The fan is mounted on the water pump shaft. It is driven by the same belt that drives the pump and dynamo. The purpose of radiator is to provide strong draft of air through the radiatorto improve engine cooling
- **Water Jacket:** Water jackets are passages cored out around the engine cylinder as well as around the valve opening
- **Temperature Gague:** It is a device that measures the coolant temperature circulating through the engine, providing the drive with a visual indication ofthe engine's operating temperature and alerting them if it become too hot or cold, allowing for timely adjustments to prevent optential damage.
- **Hosepipe:** Those one essential for connecting various components of the cooling system, allowing coolant to flow between them.

Direct or Non-return System

The direct or non-return cooling system is a basic approach used in certain internal combustion engines. In this setup, water is circulated directly from the engine to the radiator using a centrifugal pump, typically powered by a belt linked to the engine's crankshaft. Unlike systems equipped with a thermostat, this design lacks any mechanism to control the flow of coolant, leading to a continuous flow from the engine to the radiator and back again. Although this system is relatively simple and economical, it may not offer accurate temperature regulation, particularly in cooler climates.

Advantages of the Direct or Non-return Cooling System:

This system offers several benefits:

- Its simple design makes it affordable to produce and easy to maintain.
- With fewer components and mechanisms, the chances of failure are significantly reduced, enhancing reliability.
- The lack of a thermostat and complex control mechanisms lowers the initial cost of implementation.
- Continuous coolant circulation can help maintain consistent engine temperatures under specific conditions.

Disadvantages of the Direct or Non-return Cooling System

However, the system has certain drawbacks:

- The absence of a thermostat to regulate coolant flow can result in overcooling during cold weather, affecting engine efficiency and increasing emissions.
- Inconsistent coolant flow may lower fuel efficiency, especially during the engine's warm-up phase when it needs to reach its ideal operating temperature.
- Without a thermostat, precise temperature control is not possible, which can lead to overheating in high-temperature environments.

COMPONENTS OF COOLING SYSTEM IN IC ENGINE

The Cooling system of IC Engine has several components:

Radiator

The radiator consists of aluminum tubes and strips configured in a zigzag arrangement. It is essential for engine cooling by facilitating the passage of high-temperature coolant through its tubes. As the heated fluid traverses the radiator, it conveys its thermal energy to the ambient air, which is then released into the atmosphere.

Cooling Fan

The cooling fan, situated near the radiator and adjacent to the engine, functions to prevent contact with moving components and to direct airflow. It circulates air over the radiator to expel heat from the heated coolant during engine operation, facilitating the reduction of the radiator's temperature. Contemporary electric fans are governed by computer systems that react to engine temperature through a specialized sensor.

Pressure Cap and Reserve Tank

Modern radiators are equipped with pressure caps designed to handle the expansion of pressurized coolant. These caps include a spring-loaded valve calibrated to a specific pressure rating (measured in pounds per square inch or psi). When the pressure surpasses this limit, the cap opens to release excess coolant into a reserve tank. Usually constructed from plastic, the reserve tank serves as a storage space for the overflow and offers a visual gauge for monitoring the coolant's temperature.

Water Pump

The water pump plays a vital role in ensuring the continuous circulation of coolant throughout the engine. Typically located at the engine's front, it is commonly made of cast iron or cast aluminum. It is equipped with an impeller blade that drives the coolant through the system, enabling efficient temperature regulation.

Thermostat

The thermostat functions as a temperature-sensitive valve, regulating the flow of engine coolant. It stays closed when the coolant temperature is below a designated level. Once the coolant reaches a specified threshold, the thermostat opens, enabling the coolant to circulate into the radiator for cooling.

Heating Core

The heater core functions to supply heat to the vehicle's interior when required. Resembling a radiator, it is linked to the water pump by rubber hoses. A fan blower circulates air through the heater core, facilitating the passage of heat from the heated coolant to the vehicle's interior.

Hoses

Hoses provide the comprehensive circulation of coolant within the cooling system, ensuring its flow from the radiator to several engine components and back. The top and lower radiator hoses are the most conspicuous, characterized by their greater dimensions.

Bypass System

The bypass system becomes active when the engine coolant reaches a high enough temperature to open the thermostat. It permits the coolant to bypass the radiator and flow directly back to the engine, ensuring the coolant temperature remains optimal. These systems may utilize rubber hoses or fixed steel tubes.

Cylinder Head Gaskets and Intake Manifold Gaskets

These gaskets are essential for sealing the mating surfaces of the combustion chamber, thereby preventing coolant and oil leaks into or out of the engine. Despite meticulous machining and stringent tolerances, gaskets provide a dependable seal for the coolant channels.

Freeze Plugs

Freeze plugs, created by molding specialized sand and molten metal, are engineered to conform to the contours of the engine block's coolant passageways. Their principal function is to inhibit coolant leakage from the engine. These plugs are essential for preserving the engine's integrity by preventing coolant leakage.

Requirement for a cooling system in an internal combustion engine

The function of the cooling system in an internal combustion engine is to prevent overheating by dispersing surplus heat.

- Ensure proper operating temperatures for engine components.
- Guarantee optimal combustion and comprehensive engine efficacy.
- Prolong the engine's longevity by mitigating damage caused by high heat.

Specifications for Cooling Systems in Internal Combustion Engines

Numerous solid justifications require the implementation of cooling systems in engines:

- **Rapid Heat Dissipation:** A vital feature of these systems is the prompt elimination of surplus heat when the engine overheats.
- **Temperature Regulation:** Cooling systems effectively dissipate heat from the cylinder walls, ensuring temperatures remain below designated thresholds.
- **Efficiency Optimization:** Achieving an optimal cooling balance is crucial, since excessive cooling can reduce overall engine efficiency.
- **Thermal Efficiency Preservation:** Excessive heat removal can negatively impact thermal efficiency, necessitating a cooling limitation of around 30%.

8

Working of Thermostat Valve

INTRODUCTION

A thermostat is a control valve in the cooling system that regulates water flow based on a temperature signal. It plays a vital role in maintaining the engine's optimal operating temperature. Positioned within the engine's cooling system, typically between the engine and the radiator, the thermostat is specifically engineered to determine the precise timing for opening and closing the valve. By doing so, it ensures that internal combustion engines consistently operate within the desired temperature range, promoting efficiency and preventing overheating or overcooling.

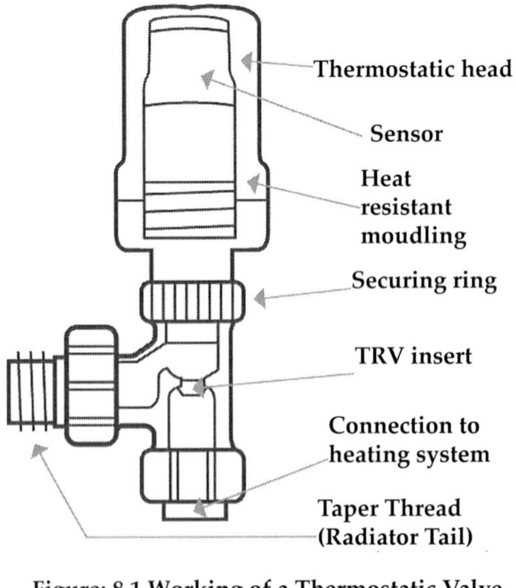

Figure: 8.1 Working of a Thermostatic Valve

WORKING OF A THERMOSTATIC VALVE

The operation of thermostatic valves involves a temperature-sensitive mechanism that regulates coolant flow:

- When the engine is cold, the thermostat valve stays closed, preventing coolant from circulating to the radiator.
- As the engine warms up during operation, the valve gradually opens, enabling the coolant to flow through the system. This helps reduce the engine's overall temperature, with the heated coolant being directed to the radiator for cooling.
- This process is continuous, as the thermostat constantly monitors the coolant temperature and adjusts the valve's opening or closing to maintain the engine within its optimal temperature range.

FUNCTION OF A THERMOSTAT IN AN ENGINE COOLING SYSTEM

The thermostat manages coolant flow to maintain the engine's ideal operating temperature in two primary states:

Closed During Cold Start

When the engine is first started, the thermostat remains closed, preventing coolant from reaching the radiator. This helps the engine warm up faster to its optimal temperature, improving fuel efficiency and reducing wear.

Open at High Temperature

Once the engine temperature reaches a specific level, typically between 190-220°F, the thermostat opens to allow coolant to flow to the radiator. This helps dissipate excess heat and prevents overheating.

CORRECT INSTALLATION OF THE THERMOSTAT

To ensure the thermostat functions properly, it must be installed correctly:

- **Installation Direction:** Thermostats are typically marked with "top" (engine side) and "bottom" (radiator side). The top should face the engine, and the bottom should face the radiator.
- **Spring End Faces Coolant Flow:** The spring end of the thermostat should face the direction of the coolant inflow to accurately sense the engine temperature and adjust accordingly.
- **Check the Sealing Gasket:** Inspect the sealing gasket around the thermostat for damage. A damaged gasket could cause coolant leaks.
- **Use a Torque Wrench for Tightening:** Tighten the bolts using the appropriate torque as specified in the equipment manual. Over-tightening can damage parts.

Incorrect installation can disrupt coolant circulation, which may lead to engine overheating or serious damage.

HOW TO TEST ENGINE THERMOSTAT?

There are three effective methods to determine if the thermostat is functioning properly:

Coolant Temperature Test

Start the engine and allow it to reach its normal operating temperature. Use a thermometer to measure the coolant temperature at the radiator inlet. If the temperature reaches the thermostat's rated value but the thermostat does not open, it may indicate a problem with the thermostat.

Manual Touch Test

After the engine has been running for some time, carefully touch the radiator's water inlet pipe (be cautious as it may be hot). If the pipe remains cold even after the engine has warmed up, the thermostat is likely not open.

Hot Water Test

Submerge the thermostat in a pot of hot water and use a thermometer to monitor the water's temperature. As the water temperature reaches the thermostat's calibrated value (usually around 195°F), the thermostat valve should gradually open. If the valve opens partially or not at all, the thermostat is malfunctioning.

Although small, the thermostat plays a crucial role in the engine's cooling system. Understanding its function, proper installation, and testing methods can help maintain your equipment, such as excavators, tractors, or loaders, in optimal condition.

For thermostat replacements, opt for high-quality parts available from the MyMROmarts store. Regularly inspect your cooling system and replace the thermostat promptly to ensure your equipment operates efficiently.

FAN

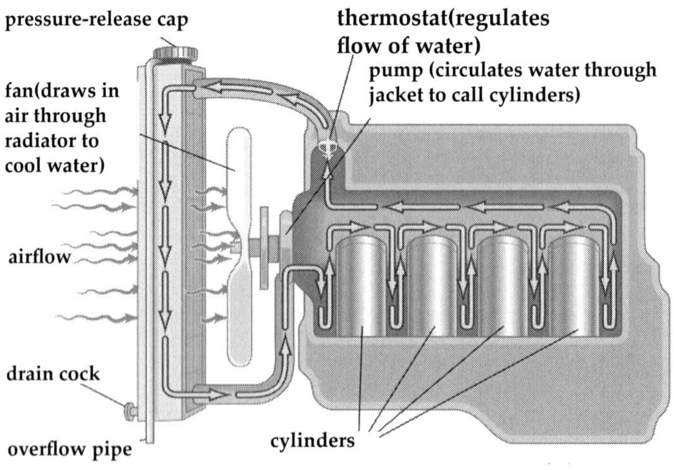

Figure: 8.2 Forced Circulation Cooling System- Water Cooled Engine

The fan is mounted on the water pump shaft and is driven by the same belt that powers both the pump and the dynamo. Its purpose is to generate a strong airflow through the radiator, enhancing the engine's cooling efficiency.

Water jackets are channels carved around the engine cylinder and valve openings to circulate coolant and regulate the engine's temperature.

DIFFERENT TYPES OF THERMOSTATIC VALVES

Thermostatic valves can be classified into two primary types based on their temperature-sensing mechanism and control:

Wax-based Thermostatic Valves

These valves rely on a wax-filled cartridge as the temperature-sensing element. The wax inside the cartridge expands or contracts in response to temperature fluctuations, triggering the valve to open or close accordingly.

Liquid-based Thermostatic Valves

A more recent innovation, liquid-based thermostatic valves use a liquid-filled sensing element that reacts more quickly to changes in temperature. This type of valve is commonly used in industrial and commercial settings, particularly in thermal management systems, due to its rapid and precise adjustment capabilities.

FUNCTIONS OF THERMOSTATIC VALVES

Here are some key functions of the thermostatic valve:

- **Temperature Regulation:** The main role of the thermostatic valve is to ensure the engine runs within its optimal temperature range, promoting peak performance.
- **Overheating Prevention:** By controlling the flow of coolant, it plays a crucial role in protecting the engine from overheating.
- **Fuel Efficiency:** The valve helps maintain the efficiency of fuel combustion by ensuring the engine operates at the ideal temperature.
- **Efficient Cold Start:** During a cold start, the thermostatic valve helps the engine warm up efficiently by keeping the coolant inside the engine rather than directing it to the radiator.

Most Common Problem of Thermostatic Valves

A common issue with thermostatic valves is getting stuck in either the closed or open position. This malfunction can significantly disrupt the engine's temperature regulation, causing it to either overheat or remain too cold. Consequently, the engine's performance can be severely compromised, affecting overall vehicle operation.

How to Check for a Stuck Thermostatic Valve?

To assess the condition of the thermostatic valve, follow these steps:

1. Begin by disconnecting the negative battery cable to safely start the testing process. Drain some coolant from the engine system to minimize any risk of spills during testing.

2. Next, remove the upper radiator hose. This will allow you to access and remove the thermostat housing that encases the thermostat. While removing it, take note of the temperature marked on the thermostat.
3. Place a thermometer in a pot of cold water to test the thermostat. Ensure the thermometer is fully submerged, then place the pot on a stove. Gradually heat the water, closely monitoring the thermometer.
4. As the water temperature reaches the marked value on the thermostat, the thermostat will open. This indicates that the thermostat is functioning correctly.

The thermostatic valve is an essential component in a car's cooling system. Regular maintenance and monitoring of the thermostat ensure the vehicle operates efficiently and can help prolong its lifespan.

GOVERNOR

A governor is a mechanical device meant to regulate the engine speed within set limits, utilized in tractors or stationary engines.

1. Sustaining a virtually uniform engine speed under varying load situations
2. Safeguarding the engine and associated equipment from elevated speeds when the load is diminished or eliminated.

Categories of Governors

1. Centrifugal governor
2. Pneumatic governor
3. Hydraulic governor

Governor Regulation

The governor is installed on an engine to maintain a constant speed; nonetheless, considerable variation in speed is noted under both full load and no load conditions. A fluctuation of approximately 100 revolutions per minute is noted between full load and no-load circumstances in a well-functioning governor. Therefore, it is feasible to adjust the governor to sustain a greater or lower speed by altering the tension of the spring. The degree of regulation is quantified as a percentage, referred to as percentage regulation. This is sometimes referred to as speed decrease. It is the fluctuation in engine speed between full load and no load conditions. It is typically represented as a percentage of the rated speed. This is given by

$$R = \frac{N_1 - N_2}{(N_1 + N_2)/2} \times 100$$

Where ,

R – % regulation,

N^1 – Speed at no load, rpm

N^2 – Speed at full load, rpm

Problem- Find the percentage regulation in a governor if speed at no load is 1600 rev/min and at full load is 1500 rev/min

Governor Hunting

Governor hunting refers to the erratic fluctuation in the speed of an engine caused by the governor overcompensating for speed variations. This phenomenon occurs when the governor causes periodic speed changes, making the engine speed increase too rapidly, then decrease just as quickly, and repeat the process. It happens when the engine accelerates quickly, prompting a sudden response from the governor, which causes the speed to drop just as fast. This cycle continues. Governor hunting may be caused by improper adjustments of the fuel pump or carburetor, incorrect idling screw settings, or excessive friction. It can also occur if the governor is too stiff or if there are obstructions preventing smooth movement of the governor components.

9

Study of Tractor/Power Tiller Engine System

INTRODUCTION

The engine of a tractor is often referred to as its heart, as it drives the machine's ability to carry out essential heavy-duty tasks in farming. As technology progresses and environmental sustainability becomes a key focus, it is increasingly important for Indian farmers in 2025 to understand the workings of tractor engines. This article aims to simplify the intricacies of tractor engines by covering their types, key specifications, performance features, maintenance, and recent innovations.

Tractors typically use diesel engines, which generate mechanical power through the compression and combustion of fuel. These engines are designed with a complex structure that ensures high power output and durability, which are vital qualities for the demanding work tractors are used for. To effectively meet these needs, tractor engines must be both powerful and built to last.

These engines function based on the principles of compression and combustion of fuel. Here is a comprehensive analysis of their attributes and significance:

- **High force:** Tractor engines must produce significant mechanical force to execute diverse agricultural chores such as plowing, tilling, transporting, and other heavy-duty operations. Diesel engines are especially appropriate for this application due to their capacity to generate greater torque at reduced speeds, which is crucial for managing substantial loads.
- **Durability:** Tractors are frequently employed in severe locations, operating for extended periods under challenging conditions. Diesel engines are designed to endure the demands of continuous operation, making them preferred for agricultural machines. Their durable construction guarantees enduring reliability and longevity.
- **Compression and Combustion:** Diesel engines operate by compressing air within the engine cylinder. Upon the injection of fuel into this highly

compressed air, it ignites spontaneously due to the elevated temperature. This combustion produces energy that powers the tractor's mechanical systems. The higher compression ratio in diesel engines enhances both efficiency and power production.

- **Fuel Efficiency:** Diesel engines exhibit superior fuel efficiency compared to gasoline engines, as they derive greater energy from an equivalent volume of fuel. This economy is particularly crucial for tractors, which frequently function for extended periods in fields where refilling may be impractical.
- **Design and Structure:** Tractor engines are engineered to meet the substantial power requirements of agricultural operations. They are designed to be durable, featuring robust components that can withstand the vibrations and stresses associated with frequent use in harsh environments. This encompasses attributes such as enlarged pistons, fortified crankshafts, and robust cooling systems.

WHAT IS THE ENGINE CONFIGURATION OF TRACTORS?

Tractors are generally equipped with diesel engines. These engines generate mechanical power by compressing and combusting fuel. Tractor engines include a sophisticated architecture and are engineered for substantial power and longevity.

Tractor engines must possess high power and durability. Consequently, the engine structures are designed accordingly. The engine comprises components like the cylinder block, cylinder head, pistons, crankshaft, camshaft, fuel system, air system, lubrication system, and cooling system.

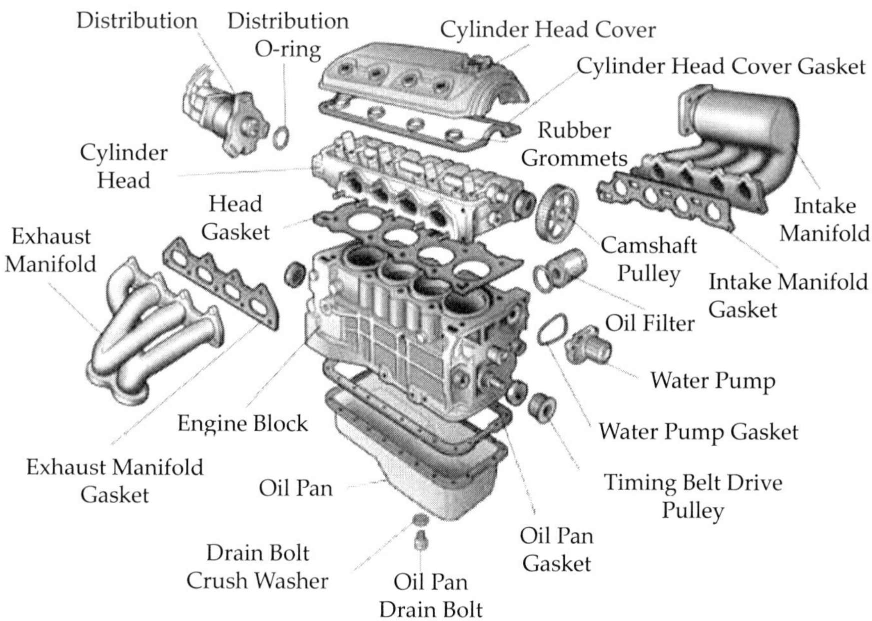

Figure: 9.1 Tractor Power Tiller Engine System

The cylinder block serves as the engine's main body, housing the engine's cylinders, which typically range from 4 to 12 in number. Inside these cylinders, pistons move, compressing the combustion chamber to generate power. The cylinder head covers the top of the cylinders and provides a seal with the cylinder head gasket, ensuring proper compression.The pistons move within the cylinders via connecting rods, which are attached to the crankshaft. This motion is converted into rotational energy by the crankshaft, transferring power to the tractor's transmission system.

To fuel the engine, the tractor's fuel system draws fuel from the tank and delivers it to the combustion chambers via the fuel pump. The air system facilitates the engine's air requirements by drawing clean air through an air filter and directing it into the cylinders.The lubrication system is essential for reducing wear and friction on engine components. It pumps clean oil to the engine parts, passing through the oil filter to ensure cleanliness. Lastly, the cooling system helps regulate the engine's temperature by cooling it with coolant or air, maintaining optimal operating conditions.

TYPES OF ENGINES USED IN TRACTORS AND POWER TILLERS

Tractor engines can be classified based on the fuel type, cylinder count, and the horsepower output.

The following are most common types of tractor engines:

Diesel Engines

Diesel engines are the most widely used type of engine in tractors due to their efficiency and power. These engines are classified based on the horsepower they produce, with most operating on the four-stroke cycle. The crankshaft of a traditional diesel engine rotates in response to speed and load changes, generating power for the tractor. Diesel engines are favored for modern tractors because they offer higher thermal efficiency compared to other internal combustion engines, making them more suitable for heavy-duty farming tasks.

Two-cylinder Engines

Tractors can be equipped with either single or two-cylinder engines, with two-cylinder engines commonly found in older models due to their simplicity and reliability. Two-stroke diesel engines, often used in smaller tractors, are popular because of their ease of use, low cost, and fuel efficiency. These engines typically have a power range between 10 and 60 horsepower and are commonly used in applications such as lawnmowers, utility vehicles, and small tractors. However, compared to four-stroke engines, two-stroke diesel engines have lower power-to-weight ratios and produce more pollutants.

Four-cylinder Engines

Four-cylinder engines are widely used in modern tractors due to their powerful performance and efficiency. Four-stroke diesel engines, the most common type in tractors, are favored for their high power output, durability, and fuel efficiency. These engines typically produce between 25 and 600 horsepower and are commonly found in heavy-duty tractors, construction equipment, and agricultural machinery. Compared to two-stroke engines, four-stroke diesel engines offer greater efficiency and generate fewer emissions, making them a more environmentally friendly option.

Six-cylinder Engines

Six-cylinder engines are few in tractors but are utilized in larger modals that require more power.

Turbocharged Diesel Engines

Turbocharged diesel engines are commonly found in high-performance tractors due to their ability to deliver increased power output and improved fuel efficiency. These engines utilize a turbocharger to enhance airflow into the engine, which in turn boosts its power. With a power range of 75 to 600 horsepower, turbocharged diesel engines are typically used in heavy-duty tractors, construction equipment, and agricultural machinery, where high performance and low fuel consumption are essential.

Intercooled Diesel Engines

Intercooled diesel engines are designed for tractors that demand high power output and excellent fuel efficiency. These engines are equipped with an intercooler, which cools the compressed air before it enters the engine, leading to improved power output and fuel efficiency. With a power range of 100 to 600 horsepower, intercooled diesel engines are commonly used in heavy-duty tractors, construction equipment, and agricultural machinery, where both performance and fuel economy are crucial.

Common Rail Diesel Engines

Common rail diesel engines are commonly used in tractors that demand high fuel efficiency, low emissions, and substantial power output. These engines feature a high-pressure fuel injection system that delivers fuel to the engine's cylinders, enhancing fuel efficiency and minimizing emissions. Common rail diesel engines typically have a power range of 100 to 600 horsepower and are utilized in heavy-duty tractors, construction machinery, and agricultural equipment. Consequently, farm tractors rely on various off-highway diesel engines tailored to specific power and application needs.

Gasoline engines

Gasoline engines are frequently utilized in smaller tractors and are recogazized for their dependatrility.

Gasoline engines are less common in tractors but are still found in some models, especially in smaller or garden tractors.

- Key Features:
  - **Initial Cost:** Generally, gasoline engines are less expensive to produce and purchase.
  - **Weight:** They are lighter than diesel engines, which can be advantageous for smaller tractors.

They provide numerous benefits compared to gasoline engines, including:

- **Fuel efficiency:** Diesel engines are recognized for their superior fuel efficiency, allowing tractors equipped with diesel engines to traverse greater distances with reduced fuel use.

This results in cost reductions for farmers and other users reliant on tractors for their tasks.

- **Elevated torque output:** Diesel engines generate substantial torque at low RPMs, rendering them suitable for demanding jobs such as plowing fields or towing significant loads.

The higher torque output enables tractors to ascend steep inclines and navigate rugged terrain effortlessly.

- **Durability:** Diesel engines are engineered for extensive operation and possess longer lifespans compared to gasoline engines.

Diesel engines are anticipated to have a longer lifespan and necessitate fewer repairs compared to engines fuelled by other fuels.

- **Minimal upkeep:** Diesel engines necessitate less maintenance compared to gasoline engines due to their more straightforward construction and absence of spark plugs.

Consequently, farmers and other users can allocate more time to productive activities rather than maintenance tasks.

- **Fuel availability:** Diesel fuel is readily accessible and may be stored for extended durations without deterioration.

Consequently, farmers and other users will have the capacity to accumulate fuel during periods of low demand and utilize it when necessary.

- **Safety:** Diesel engines exhibit lower flammability compared to gasoline engines, rendering them safer for utilization in agricultural and industrial environments.

Diesel fuel possesses a higher flash point than gasoline, indicating a reduced likelihood of ignition in the occurrence of a leak or spill.

- **Versatility:** Diesel engines are versatile and can be utilized in a range of applications, from small garden tractors to large industrial machines used in mining and construction. Their adaptability makes them a popular choice for farmers, contractors, and others who require reliable, powerful machinery. Diesel engines offer several advantages, including better fuel efficiency, high torque, durability, low maintenance, readily available fuel, safety, and versatility. These benefits contribute to the widespread preference for diesel engines among tractor users.

In summary, the most common types of tractor engines are gasoline and diesel, and they can be classified by the number of cylinders and the horsepower they produce.

Figure: 9.2 Diesel Engines

Operational Mechanism of Tractor/Power Tiller Engines

The engine system functions on either a four-stroke cycle or a two-stroke cycle, contingent upon the engine's design. The four-stroke cycle is predominantly utilized in agricultural engines because to its superior fuel efficiency and durability.

Four-Stroke Cycle (Prevalent in Tractor Engines)

- **Intake Stroke:** The intake valve opens, and the piston descends, drawing in air (and fuel in diesel engines).
- **Compression Stroke:** The piston ascends, compressing the air-fuel mixture. In diesel engines, solely air undergoes compression.
- **Power Stroke:** At the apex of compression, fuel is injected (in diesel engines) and ignites due to the thermal energy of compressed air. The combustion propels the piston downward, producing power.
- **Exhaust Stroke:** The exhaust valve opens, and the piston ascends, ejecting the combustion gasses from the cylinder.

Two-Stroke Cycle (Less Common but Utilized in Smaller Engines)

Integrates the intake and compression strokes into a single action, as well as the exhaust and power strokes into another. It is less complex but exhibits inferior fuel efficiency and increased pollution.

Figure: 9.3 Two-stroke Tractor Engine

Key Specifications of Tractor Engine

Comprehending the specs of a tractor engine aids in selecting the appropriate model for your agricultural needs.

Cubic Capacity (CC)

Tractor engines typically vary from 1,000 cc to 4,500 cc, with larger capacities appropriate for heavy-duty usage. The cubic capacity influences the engine's power production and fuel efficiency.

Cooling Mechanisms

- **Air-Cooled:** Appropriate for compact tractors and optimal for areas with temperate climates.
- **Liquid-Cooled:** Frequently utilized in high-capacity engines to guarantee excellent performance during demanding operations, particularly in elevated temperatures.

Horsepower (HP)

Ascertains the engine's power. A tractor engine of 20-40 HP is suitable for small farms, however larger farms may necessitate engines above 50 HP to manage vast operations well.

Performance Attributes

The performance of tractor engines is essential, directly influencing efficiency and productivity in agricultural operations.

- **Torque Output:** Elevated torque at low RPM is crucial for activities such as plowing, hauling, and utilizing heavy equipment. Contemporary tractor engines provide torque ranging from 100 Nm to 400 Nm, allowing them to manage substantial loads effortlessly.
- **Fuel Efficiency:** Contemporary tractors exhibit substantial enhancements, providing approximately 30-40% superior fuel economy compared to their predecessors. This guarantees reduced operational expenses, an essential consideration for farmers.
- **Durability:** Engines are now constructed with superior materials and design elements to endure extended operation in challenging agricultural environments, guaranteeing reliability.

Emission Regulations

In response to increasing environmental concerns, India has implemented stringent emission restrictions for agricultural machinery.

- **BS-VI Compliance:** Tractors now conform to Bharat Stage VI regulations, minimizing detrimental emissions and harmonizing with international standards.

Advanced Technologies

- **Common Rail Direct Injection (CRDI):** Improves fuel atomization for optimal combustion, resulting in increased power production and reduced emissions.
- **Exhaust After-Treatment Systems:** Filter pollutants to ensure adherence to environmental rules and enhance air quality.

Maintenance Considerations

Consistent maintenance guarantees the durability and dependable operation of a tractor engine. Below are essential maintenance tasks:

- **Oil Changes:** Substitute engine oil after every 100 hours of operation to avert deterioration. Pristine oil guarantees seamless functionality and safeguards interior components.
- Air and fuel filters should be cleaned or replaced routinely to ensure optimal engine efficiency and to prevent pollutants from infiltrating the engine.
- **Coolant Levels:** Regularly assess and replenish coolant to avert overheating, particularly during extended operational periods in the field.
- Examine belts and hoses for deterioration and repair them as necessary to prevent malfunctions and maintain optimal engine performance.
- **Service Schedule:** Adhere to the manufacturer's prescribed service schedule, which generally entails thorough inspections every 500 hours of operation.

Which Types of Engines are Utilized in Tractors?

Most tractors are equipped with diesel engines featuring Direct Injection systems, and they are classified based on the horsepower they produce. Modern engines range from 12 horsepower to 120 horsepower, providing a variety of options for different agricultural needs. The majority of these engines operate on the four-stroke cycle, with the crankshaft rotating in response to speed and load to generate power.Tractors typically come with 2 to 4-cylinder engines, although in India, it's common to find tractors with 1 to 4 cylinders. Additionally, while diesel remains the primary fuel used for tractors in India, some newer models are powered by electric motors, reflecting an emerging trend toward more environmentally friendly power sources.

What is the Operational Mechnism of a Tractor Engine in relation to other Implements?

A tractor operater using a power take-off. It employs a grooves connected point when attached to a PTO driven Implement to rotate the shaft to a set rate. Also, tractor engines are powered by diesel and have a transmission system, often comprising mostly 8-speed or 10-speed gearbox for transmitting the power to work.

Stroke-Bore Ratio

A French engineer established several key principles for internal combustion (IC) engines that have since guided engine design: the combustion chamber should have the smallest possible surface-to-volume ratio, the expansion process should be as rapid as possible, and compression at the start of expansion should be maximized. These principles emphasize the importance of minimizing heat loss and maximizing efficiency.

The stroke-bore ratio is an important design consideration in engines. A higher compression ratio is more feasible with smaller bores, improving thermal efficiency. However, a larger stroke-bore ratio increases the surface area relative to volume, which leads to more heat being transferred to the combustion chamber walls, thus reducing efficiency. Shorter stroke engines, which have less piston travel, also experience less friction. High-speed engines benefit from a smaller stroke-bore ratio, improving volumetric efficiency by allowing for larger or multiple valves.

In 1982, tractors tested at Nebraska had an average stroke-bore ratio of 1.01, down from 1.29 in 1920. By 2001, the average stroke-bore ratio of tractor engines in India was around 1.33. There is a clear trend toward further reducing this ratio in modern designs. Recent tractors, such as the Swaraj 744 and some models in the Escorts Farmtrac range, have a stroke-bore ratio of 1 or even lower, with some models as low as 0.95. This reduction helps achieve higher engine speeds, more compact engine sizes, and reduced vibration. Additionally, turbocharging technology has been introduced in Indian tractors to further enhance performance.

Specific Fuel Consumption (SFC)

Fuel efficiency is a critical consideration in farm economics, with tractors that have the lowest specific fuel consumption (SFC) being the most efficient. Manufacturers continually strive to reduce the SFC of their newer tractor models. Among the six major tractor manufacturers in India, the Mahindra 575DI stands out for having the lowest SFC at maximum power, registering 233 g/kWh.

An important trend in tractor development in India is the reduction of the weight-to-power ratio, which results in tractors that can travel faster. However, this improvement in speed has not been matched by corresponding advancements in traction, making it crucial for further efforts to enhance the traction capabilities of tractors. Indian manufacturers have made significant progress in improving the weight-to-power ratio of their prime movers. This has been achieved through slimming processes on castings using advanced CAD/CAM designs and by modifying designs to accommodate higher brake mean effective pressures (BMEP), which now reach peak values close to 23 bars.

Compression Ratio

The thermal efficiency of an engine does not attain the theoretical efficiency of an Otto cycle engine as specified by

$$\text{Efficiency} = 1 - \frac{1}{r^{n-1}}$$

Where r is the compression ratio.

n = Cp/Cv (Ratio of specific heats of air at constant pressure and constant volume)

Thus, it is evident that the efficiency of an ideal Otto cycle rises with an increase in the compression ratio.

The real efficiency is significantly lower than the theoretical efficiency for numerous reasons. Several inconsistencies can be attributed to the following factors

1. The compression and expansion of gases is not adiabatic.
2. Frictional losses occur within the engine.
3. Work is necessary to intake and expel the gasses.
4. For a naturally aspirated engine, the volumetric efficiency will be below 100%.
5. Complete combustion is not achieved.

Significant alterations in the compression ratio of diesel engines are impractical. The compression ratios of most diesel engines range from 16:1 to 22:1.x

Design of Combustion Chamber

The shape and size of the combustion chamber in diesel engines significantly influence the mixing of fuel and air, as well as fuel detonation. Typically, the design aims to create turbulence within the mixture to improve combustion efficiency and enhance flame propagation velocity, which reduces combustion time and improves antiknock characteristics. A short and compact combustion chamber generally results in a quicker combustion process, further enhancing the antiknock properties of the fuel.

There are several variations in the design of combustion chambers used in tractor diesel engines, with differences arising from the various methods of initiating combustion within the cylinder. Since the fuel-to-air ratio is low when fuel is initially injected, ignition doesn't occur until the ratio reaches the combustible range. If fuel were directly injected into the combustion chamber, there would be a prolonged delay before ignition, leading to very high initial pressures.

There are four common types of combustion chambers used in diesel engines: direct injection (DI), pre-chamber, swirl or turbulence, and auxiliary combustion chambers. The most commonly used method in tractor engines is direct injection (DI), where a concave piston head is employed. This method, introduced by M&M Ltd., helps increase thermal efficiency and reduces fuel consumption. DI engines feature fine atomization of fuel in the combustion chamber, leading to better combustion.

Other tractor engine manufacturers, like HMT and Eicher, use a turbulence or swirl combustion chamber design. In these engines, the swirl action of the fuel-air mixture promotes better mixing and combustion, which increases the mean effective pressure and improves engine efficiency. These engines are classified as indirect injection (IDI) engines and are known for being quieter than DI engines. IDI engines can also operate on fuel with a lower cetane number. The swirl chamber, which contains 50 to 90% of the compressed volume at top dead center (TDC), aids in the efficient combustion of the fuel-air mixture.

Eicher Ltd. refers to these IDI engines as Ricardo-type engines, in honor of the engineer Ricardo, who first used this design. The improved swirl control in these engines contributes to lower noise levels, making them more efficient and environmentally friendly.

How Do Tractor Engines Work?

Diesel engines are the dominant power source for modern tractors due to their superior thermal efficiency compared to other internal combustion engines. The high thermal efficiency of diesel engines is primarily a result of their extremely high compression ratio. Unlike gasoline engines that rely on spark plugs for ignition, diesel engines use highly compressed hot air to ignite the fuel.

The process begins when air is drawn into the combustion chamber and compressed to a pressure of up to 580 PSI. This compression heats the air to temperatures as high as 1022°C. At this critical temperature, the fuel injector sprays finely atomized fuel droplets into the combustion chamber. These tiny fuel droplets mix with the hot, compressed air, leading to ignition. The combustion generates expanding gases, which push the piston downward. This piston movement is then converted into rotational motion via the crankshaft, powering the tractor. This method of combustion provides greater efficiency and is well-suited for the heavy-duty tasks performed by tractors.

Best Engines for Tractors

Many tractor manufacturers produce their own engines, while some source their engines from external companies. For example, Mahindra and Massey Ferguson tractors use engines manufactured by Mitsubishi. These engines are known for their durability and efficiency, contributing to the tractors' overall performance.

Cost of Rebuilding a Tractor Engine

When a tractor's engine fails, rebuilding it can be expensive. However, it is generally more affordable to repair or rebuild the engine than to purchase a new one. By opting for repairs, you can save up to 50% of the cost of a new engine, making it a cost-effective choice for many tractor owners.

How to Handle a Tractor to Keep the Engine Safe

To ensure your tractor's engine remains in optimal condition, avoid leaving it in muddy areas, as this can damage the tractor's body and components. When using the hitch system, raise it slowly to prevent undue stress on the engine. Additionally, always apply the emergency brakes and turn off the transmission when the engine is not in use. Regularly check and maintain the engine to enhance its efficiency and extend its lifespan.

HOW IS THE OPERATOR RESPONSIBLE FOR ENGINE EFFICIENCY?

A tractor operator is a skilled individual who handles the tractor for a variety of tasks, including transporting goods from fields to warehouses, conducting farming operations such as tillage, ploughing, reaping, and threshing, and performing commercial activities like transporting building materials. Apart from operating the tractor, the operator is responsible for maintaining the tractor's functionality and ensuring its safe use. A key part of this is keeping the tractor engine in good condition through regular maintenance practices.

By understanding tractor engine operations, you can ensure the longevity and efficiency of your equipment. Additionally, Tractor Junction offers valuable information on tractors and their components, providing essential details without taking up much of your time, helping you stay informed and proactive in maintaining your tractor for optimal performance.

Advancements in Tractor Engine Technology

Modern technology is driving the evolution of tractor engines to enhance efficiency and sustainability.

Smart Engine Technologies

- **Integration of Internet of Things (IoT):** Real-time surveillance of engine performance enables farmers to mitigate possible problems prior to their escalation. Sensors deliver information regarding characteristics such as temperature, oil pressure, and fuel levels.

- **Advantages:** Improves operating efficiency, minimizes downtime, and facilitates predictive maintenance.

Hybrid Systems

- **Diesel-Electric Hybrid Vehicles:** The integration of diesel and electric power decreases fuel consumption and pollution. Hybrid engine tractors offer versatility for various operational requirements, optimizing power and efficiency.
- **Present Condition:** This idea, albeit not prevalent, is being investigated by prominent manufacturers to serve ecologically aware farmers.

Alternative Fuels

Manufacturers are evaluating engines that operate on compressed natural gas (CNG) or biogas. These options are more economical and generate reduced emissions, rendering them a sustainable option for the future.

- **Government Assistance:** Subsidies and incentives for the use of alternative fuel technologies are motivating farmers to investigate these choices.

Maintenance and Care for Tractor and Power Tiller Engines

Consistent maintenance is crucial for the durability and efficiency of tractor engines. Essential maintenance activities comprise:

- Oil Change: Routine oil changes mitigate engine wear and overheating.
- Replacement of Air and Fuel Filters: Guarantees the engine is supplied with unpolluted air and fuel.
- Coolant Inspection: To prevent overheating, verify that the coolant level is enough.
- Valve and Timing Adjustment: Guarantees optimal engine performance.
- Battery Maintenance: Consistently inspect the battery's charge and terminals.

10

Power Transmission System of Tractor/ Power Tiller

INTRODUCTION

Transmission is a mechanism designed to reduce speed, consisting of multiple gears. It functions as a series of gears and shafts that transmit engine power to the tractor wheels. This system includes various components that enable both forward and reverse movement of the tractor, adapting to different field conditions. The entire route through which power flows from the engine to the wheels is referred to as the power train.

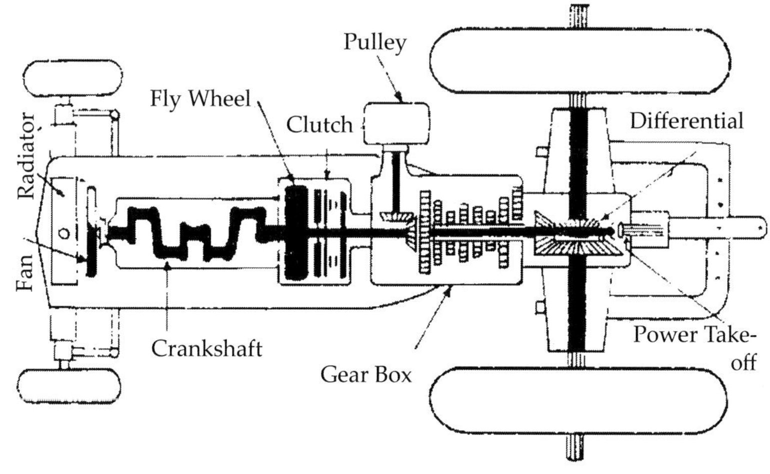

Figure: 10.1 Power Transmission System of Tractor

FUNCTIONS OF THE POWER TRANSMISSION SYSTEM

1. To transfer power from the engine to the tractor's rear wheels.
2. To provide reduced speeds to the rear wheels as required.

3. To adjust the ratio between wheel speed and engine speed to match varying field conditions.
4. To facilitate power transmission through a right-angle drive, as the crankshaft and rear axle are typically positioned at right angles to each other.

The power transmission system is composed of several components: clutch, transmission gears, differential, final drive, rear axle, and rear wheels. Together, these components ensure the effective transmission of power.

DIFFERENT TYPES OF CLUTCHES IN TRACTOR

Clutch is a device, used to connect and disconnect the tractor engine from the transmission gears and drive wheels. Clutch transmits power by means of friction between driving members and driven members.

Necessity of Clutch

The clutch in a tractor serves several critical purposes, making it an essential component of the tractor's transmission system:

- **Engine Cranking:** To start the engine, it must be disconnected from the rest of the transmission unit. This is accomplished by the clutch, allowing the engine to be cranked easily. After the engine starts, the clutch is engaged to transmit power from the engine to the gearbox.
- **Gear Changes:** The clutch ensures that the gearbox is disengaged from the engine's power when changing gears. This prevents damage to the gear teeth and ensures smooth gear engagement, as the power flow to the gearbox is interrupted during gear shifting.
- **Belt Pulley Operation:** When the tractor's belt pulley is in operation in the field, it needs to be stopped without shutting down the engine. The clutch provides this functionality by disengaging the pulley, allowing it to stop independently while the engine continues running.

Overall, the clutch is crucial for efficient tractor operation, preventing damage to components, facilitating gear changes, and providing flexibility in power management.

Essential features of a good clutch

1. Its hould have good ability of taking load without dragging and chattering.
2. It should have higher capacity to transmit maximum power without slipping.
3. Friction surface should be highly resistant to heat effect.
4. The control by handl ever or pedall ever should be easy.

Types of clutch

Clutch esaremainly of three types:

1. Frictionclutch
2. Dogclutch
3. Fluid coupling.

Friction clutch is most popular in four wheel tractors. Fluid clutch is alsoused in some tractors thesedays.Dog clutch is mostly used in power tillers. Friction clutch may be subdivided into three classes: (a) Single plate clutch or single disc clutch (b) Multiple plate clutch or multiple disc clutch (c) Cone clutch.

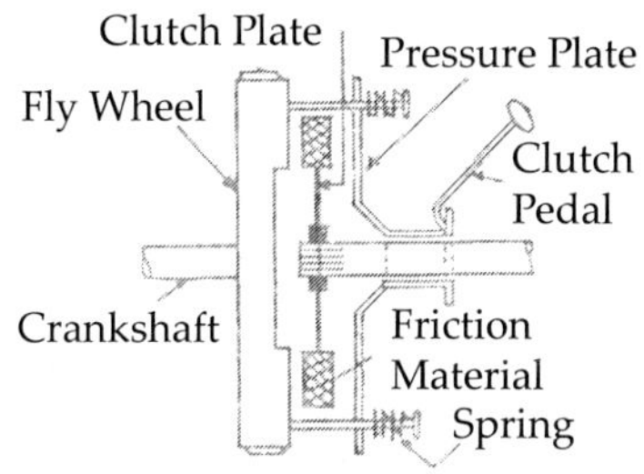

Figure: 10.2 Single Plate Clutch

GEARS IN TRACTOR

Speed varies according to the field requirements and so a number of gearratios are provided to suit the varying conditions. Gears are usually made of alloysteels. As the tractor has to transmit heavy to rqueall the time,best quality lubricants free from sediments,grit,alkali and moisture,is used for lubrication purpose. SAE 90 oil is generally recommended for gear box.

A tractor runs at high speed, but the rear wheel of the tractor requires power at low speed and high torque. That's why it becomes essential to reduce the engine speeed and increase the torque available at the rear wheel of the tractor because

$$\text{Power. kW} = \frac{2\,\pi\,NT}{60\times1000}$$

Where,

T = torque in Newton–meter

N = speed in rev/min

If engine power is constant, it is obvious that for higher torque at wheels , low speed is required and vice versa.. So gear box is fitted between engine and rear wheels for variable speed and torque.

A gear is a toothed mechanical component designed to transfer torque between rotating parts, using its teeth (or cogs) to engage with another gear or toothed element. The teeth are precisely shaped to minimize wear, vibration, and noise, while maximizing the efficiency of power transmission.Gears of different sizes are often used in pairs, providing a mechanical advantage. This allows the driving gear's torque to either produce a larger torque in the driven gear at lower speeds or a smaller torque at higher speeds. The larger gear in the pair is called the wheel, while the smaller one is referred to as the pinion. This principle is fundamental in systems like automobile transmissions, where different gear ratios allow for varying mechanical advantages depending on the driving conditions. It's important to note that a gearbox does not amplify power or function as a servomechanism. According to the conservation of energy, the power output from the driven gear will never exceed the power applied to the driving gear. There will always be some power loss due to friction within the system, which reduces efficiency slightly.

Differential

The differential of a tractor consists of a planetary gear system with four bevel gears: two side gears and two pinions. It also includes a pair of bevel gears—comprising a pinion and a crown wheel—fitted at right angles to each other to transmit power from the gearbox to the rear axles. The input power is delivered to the planet carrier attached to the crown wheel, while the output power goes to the side gears mounted on the counter shafts, driving the rear wheels. The differential serves as a speed reduction mechanism, transmitting power at a 90° angle. It divides the power equally between the wheels and allows one wheel to move faster than the other while turning.

When the tractor moves straight ahead, the bevel pinions of the differential do not rotate on their carrier shafts, and both side gears rotate at the same speed. A differential lock is also included to prevent one wheel from spinning excessively when the resistance to one wheel is different from the other. For example, if one wheel is stuck in mud or loose soil, the other wheel, on solid ground, would not move while the spinning wheel rotates. The differential lock allows both wheels to rotate at the same speed and apply equal torque, solving this issue.

The differential unit enables the rear wheels of the tractor to rotate at different speeds when turning. The inner wheel, which travels a shorter distance on a curve, requires less power than the outer wheel. The differential unit ensures this by allowing one wheel to rotate faster than the other during turns.

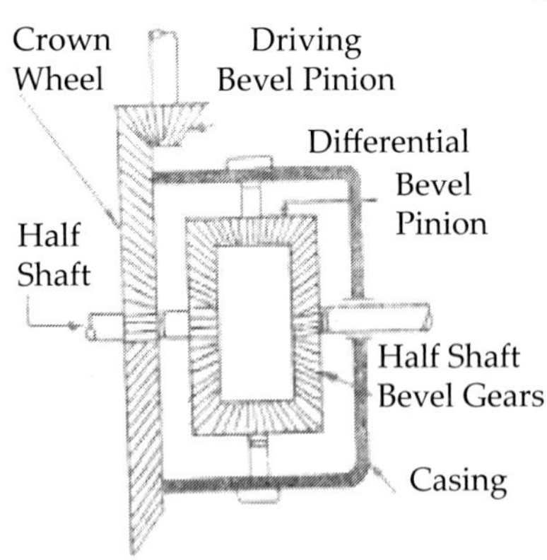

Figure: 10.3 Tractor Differential Unit

Differential Lock

Differential lock is a device to join both halfaxles of the tractor sothatevenif one wheel is under less resistance, the tractor comes out from themudetcas both wheels move with the same speed and apply equal traction.

Design Considerations for Bevel Gears in a Differential

When designing a bevel gear drive for a differential, the following factors should be considered:

1. The gear must possess adequate strength to with stand starting torques and dynamic running conditions without failure.
2. The teeth should exhibit excellent wear resistance to ensure a long service life for the gear.
3. The chosen material should have superior wear characteristics to enhance the gear's durability.
4. The drive should be compact and efficiently designed to optimize performance.
5. Proper lubrication arrangements must be implemented to ensure smooth operation.

Final Drive

The final drive is a gear reduction unit located in the powertrain between the differential and the drive wheels. It transmits the power to the rear axle and wheels of the tractor. Rather than attaching the rear wheels directly to the half shafts, the drive is transmitted through a set of spur gears. Each half shaft ends in a small gear that meshes with a larger gear, known as the bull gear. The bull gear is mounted on a shaft that holds the tractor's rear wheel. The mechanism responsible for the final speed reduction, specifically for the rear wheels of the tractor, is called the final drive mechanism.

Spur Gears

These are Flat with teeth projecting radially in the plane of the wheel, "straight-cut gears". These gears can be fitted only to parallel axles.

Figure: 10.4 Spur Gear

Helical Gears

A refinement over spur gears The teeth are cut at an angle, allowing for more gradual, hence smoother meshing between gear wheels, eliminating the whine characteristic of straight-cut gears. - Double helical gears - Also known as herringbone gears. These gears have teeth that are 'V' shaped. Each gear in a double helical gear can be thought of as two standard, but mirror image, helical gears stacked. This cancels out the thrust since each half of the gear thrusts in the opposite direction. They can be directly interchanged with spur gears without any need for different bearings.

Figure: 10.5 Helical Gear

Bevel Gears

Bevel gears are designed to change the operating angle. Each gear is specifically paired to function with its complementary gear and cannot operate with any other. A differential gear system, comprising four bevel gears arranged in a square configuration, enables the transmission of power to two axles that rotate at different speeds, as seen in vehicles navigating a corner.

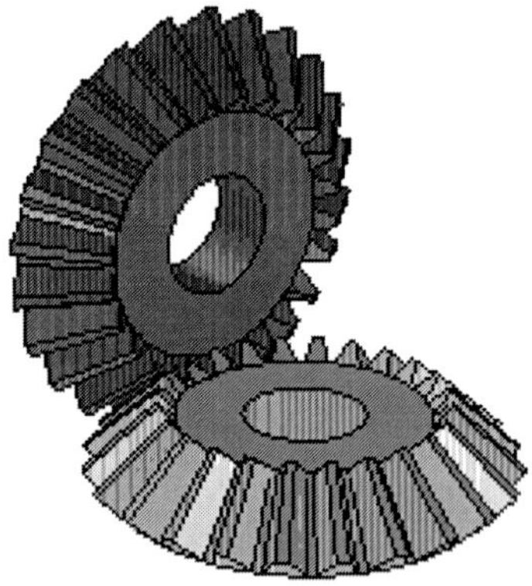

Figure: 10.6 Bevel Gear

SLIDING MESH GEAR BOX IN TRACTOR

The sliding mesh gearbox is one of the oldest and simplest designs of automotive gear systems. It uses gears that are manually shifted by the operator to change speeds. However, this method requires more effort compared to modern gearboxes, such as constant or synchromesh gearboxes. The operator must ensure the gears are aligned properly before shifting; otherwise, the gears will grind, creating a strong noise. To prevent damage to the gears, gear changes should be made carefully, especially at low speeds.

The sliding mesh gearbox is simple, cost-effective, and robust, making it suitable for high-torque applications. However, it has some disadvantages, such as being harder to operate, noisier, and more difficult to shift smoothly. As the name implies, the main shaft gear slides along the main shaft to mesh with the corresponding gear on the counter shaft. The main shaft is splined, and the main shaft gears are also splined from the inside, ensuring positive motion between them. Despite its drawbacks, this design has been used widely due to its low manufacturing cost and durability.

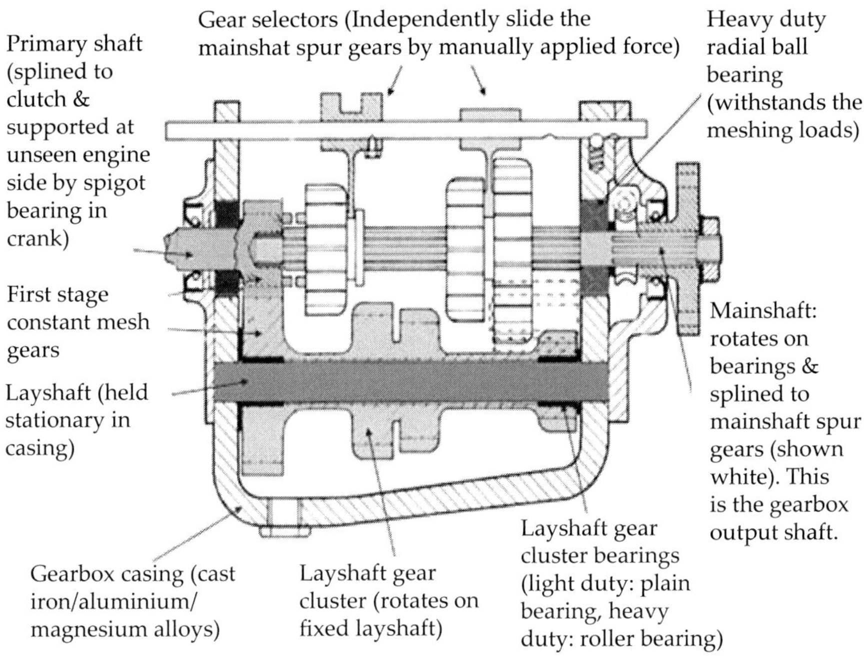

Figure: 10.7 Sliding Gear Tranmission

In a sliding mesh gearbox, only one set of gears, consisting of a main shaft gear and a countershaft gear, should be meshed at any given time. If two pairs of gears get meshed simultaneously, they will attempt to rotate the main shaft at different speeds, which could lead to the breakage of either the main shaft or the meshed gears.

To engage a particular gear, it is slid along the main shaft by its collar, and it is then made to mesh with the corresponding countershaft gear. For example, gear 4 on the main shaft corresponds to the 1st gear, gear 5 corresponds to the 2nd gear, and gear 6 corresponds to the 3rd gear. In the case of the 4th gear, the primary and secondary shafts are coupled together, which means there is no speed reduction in the gearbox when this gear is selected.

This design ensures that only one gear pair is engaged at a time, preventing damage and ensuring smooth operation. The sliding mesh gearbox is a simple yet effective design for power transmission in vehicles.

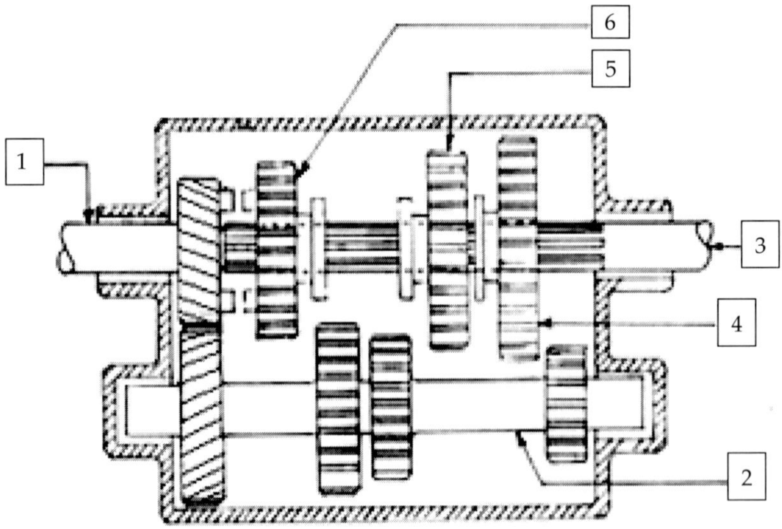

Figure: 10.8 Cross-sectional View of a Sliding Mesh Gear Box, 1. Input Drive Gear, 2. Counter Shaft, 3. Main Shaft, 4. Ist Gear, 5. IInd Gear, 6. IIIrd Gear

During the neutral position, none of the main shaft gears are engaged to the counter shaft gears. Depending on the speed-torque requirement, the gears of main shaft is slid to mesh with the respective gear on the counter shaft.

CONSTANT MESH GEAR BOX IN TRACTOR

In a constant mesh gearbox, the gears remain continuously engaged, and gear shifting is achieved by disengaging and then re-engaging the collar with the desired gear. The operator must select the required gear pair and synchronize their speeds using the clutch for smooth shifting. This design provides smoother transmission compared to a sliding mesh gearbox. Most tractors up to 50 HP now use constant mesh gearboxes, which are well-suited for Indian farming needs. Over time, a combination of constant mesh and sliding mesh gearboxes was developed, known as a partial constant mesh gearbox, offering a balanced approach to cost, complexity, and performance.

In case of constant mesh type of gear box all the gears of main shaft are constantly meshed to the corresponding gears on the counter shaft as shown in the figure below:

Similar to the sliding mesh gear box, the main shaft is splined in this case too. But since all the gears on the main shaft are meshed, these are free to rotate on the splined main shaft. To transmit power an arrangement of fixed dog clutch and sliding dog clutch is used. The arrangement is shown as in the fig below.

In a sliding mesh gearbox, the main shaft gears are able to rotate freely on the main shaft, and the fixed dog clutch is splined to the main shaft. When the dog clutch rotates, it drives the main shaft. The sliding dog clutch, which is positioned between two gears, allows for the engagement of specific gears. During the neutral position, the sliding dog clutch sits on the fixed dog clutch, and no gear is engaged. When a specific gear, such as gear A, is to be engaged, the sliding dog clutch moves left to lock the selected gear with the fixed dog clutch, allowing the gear to mesh with the corresponding gear on the countershaft.

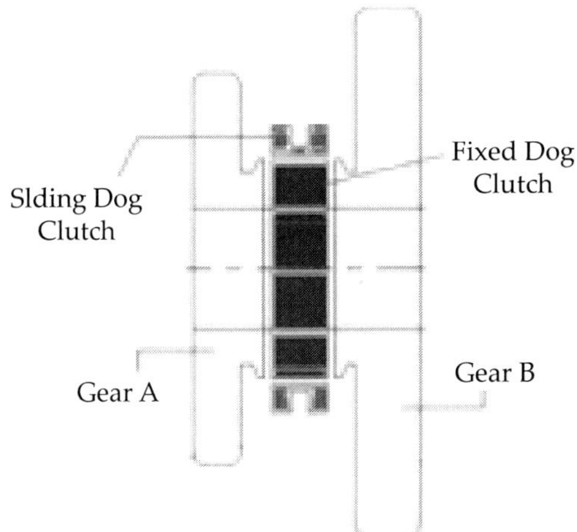

Figure: A 10.9 Transmit Power an Arrangement of Fixed Dog Clutch

Once the sliding dog clutch locks onto the gear, power is transmitted through the fixed dog clutch to the main shaft. Meanwhile, the other gears on the main shaft rotate freely. All the gears on the main shaft mesh with their corresponding countershaft gears, which are typically helical gears. These helical gears have advantages over spur gears, such as reduced noise due to point contact at the beginning of engagement, and greater strength due to more gear teeth being in contact, which allows for higher torque transmission.

When engaging a gear from the neutral position, the clutch is disengaged while the gears continue to rotate due to their momentum. The selected sliding dog clutch is then moved towards the gear that needs to be engaged. As the relative motion between the gear and the sliding dog clutch decreases, the dog clutch locks into place with the selected gear, and the clutch is gradually engaged.

In a constant mesh gearbox, the clutch is typically pressed twice when shifting from one gear to another. The first press disengages the clutch to shift to the neutral position, while the second press engages the clutch to move from neutral to the desired gear. This process is called double de-clutching and helps to ensure smoother gear transitions.

PLANETARY GEAR BOX IN TRACTOR

A planetary gearbox is a type of transmission system used to transfer power and torque from the input shaft to the output shaft. It is made up of several key components, including the sun gear, planet gears, and ring gear. These gears work in unison to achieve precise control over the speed and torque of the motor.

The sun gear is located at the center of the gearbox and serves as the primary gear. The planet gears are mounted on a carrier and orbit around the sun gear, engaging with both the sun gear and the ring gear. The ring gear, which surrounds the planet gears, also engages with the planet gears to transmit torque. This configuration allows for a compact design with high torque output and efficiency.

The planetary gearbox is known for its ability to provide a range of gear ratios and torque outputs, making it highly versatile and commonly used in applications requiring high power and compact size, such as in automotive transmissions, industrial machinery, and robotics.

How Does It Work?

A planetary gearbox functions through the synchronized interaction of several essential components: the sun gear, planet gears (or pinions), and the ring gear.

- **Sun Gear (Sun Wheel):** Positioned at the center, the sun gear is typically connected to the input or power source that drives the gearbox. It acts as the primary gear that transmits power to the system.
- **Planet Gears (Pinions):** Surrounding the sun gear are multiple planet gears, which are mounted on a rotating carrier or arm. These gears mesh with both the sun gear and the ring gear, facilitating the transmission of torque.
- **Ring Gear (Annulus):** The ring gear, or annulus, encircles the planet gears and meshes with them. It has internal teeth and is often stationary or fixed within the gearbox system.
- **Gear Interaction:** The planet gears simultaneously mesh with both the sun gear and the ring gear, creating torque through the rotation of the sun gear. This interaction ensures efficient power transfer and motion.
- **Rotation Direction:** The planet gears exhibit dual motion—rotating around their own axes while also orbiting around the sun gear. This combined rotational and orbital motion is essential for torque transmission.
- **Output:** The output of the gearbox is typically derived from the planet carrier or the sun gear, depending on the specific configuration and application needs.

The configuration of the planetary gearbox, including the number of teeth on each gear, determines the gear ratios and torque distribution. By adjusting these parameters, different gear ratios can be achieved, which allows for greater flexibility in meeting particular requirements.

Overall, planetary gearboxes offer high torque density, efficient power conversion, and compact designs, making them highly favored in applications such as automotive transmissions, robotics, and industrial machinery.

Key Components of Planetary Gearbox

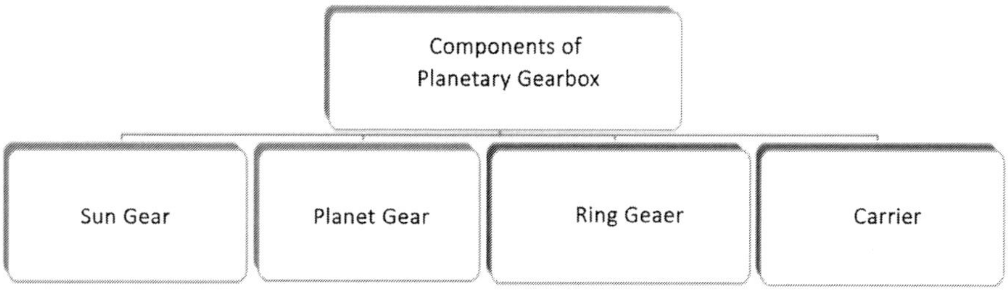

The Important components of Planetarygearbox are as follows:

Sun Gear

It is an essential component of the planetary gearbox and plays an important role in the function of the gear system. This gear is located at the centre of the gear system and is surrounded by multiple gears. This gear is smaller in size compared to the other gears as it meshes with the teeth of the gears for smooth and efficient power transmission.

Its main function is to control the speed and direction of the shaft and act as a central point of the gearbox.

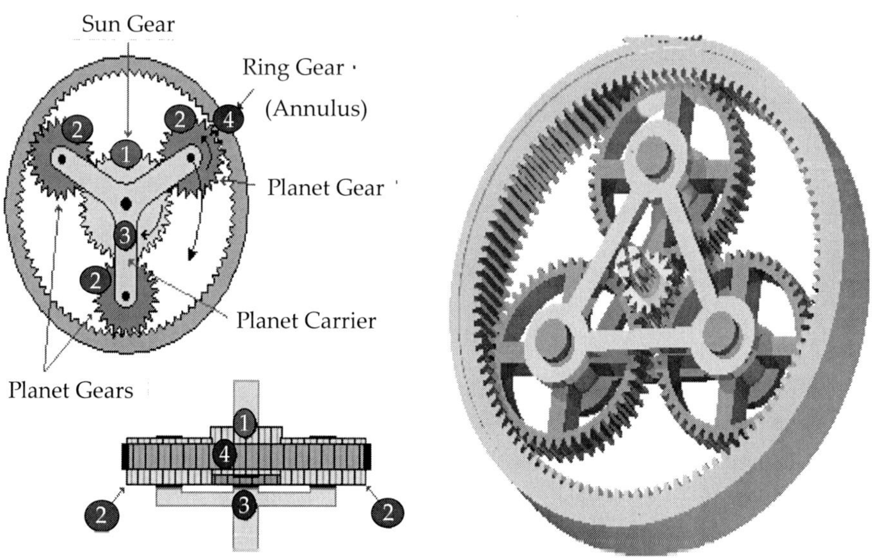

Figure: 10.10 Planetary Gear Unit

Planet Gear

An important function of planet gear is to provide gear reduction. It has several teeth that surround its circumference. These teeth mesh with the teeth of the sun gear and cause the rotation of the output shaft.

The ratio of the number of teeth of planetary gear and sun gear determines the gear reduction ratio. This ratio is important to achieve high torque at a small size as compared to other gear systems.

Ring Gear

The ring gear is a large circular gear that acts as a stationary component of a planetary gearbox. It is made up of steel and is responsible for transmitting the power from the input gear to the output shaft.

The design of the ring gear contributes to the reliability and durability of the gearbox. Its high-strengthened steel construction helps the gearbox to withstand in high torque and shock load conditions and makes it suitable for heavy-duty applications.

Carrier

- It is another important component of the planetary gearbox. It is a cylindrical structure with holes that hold the planet gear and allow it to rotate around the sun gear.
- It not only holds the planet gear but also helps in evenly distributing the load on other gears.
- It is a crucial component for the smooth and efficient operation of the gearbox without a single gear burnt with load.

Other Components

Input Shaft

It receives the power from the motor and transfers it to the gears of the planetary gearbox to determine the output speed and torque of the gearbox. It also helps to evenly distribute the load among all the gears

Output Shaft

It is located at the end of the planetary gearbox and connected with the final drive or gears. It is made up of high-strength materials to ensure smooth operation and efficient power transmission.

Advantages

- **No Backlash:** Gearing has nearly no backlash and is quite accurate.
- **Modular:** The majority of planetary phases can be stacked since they are modular.
- **Shaft:** Coaxial configuration means that there is no offset in the output shaft.
- **Gear:** increased gear life at equivalent loads.
- **High Power Density:** The burden is distributed among many planets.
- **Size:** Planetary gears are lightweight and small, making them simple to employ in a variety of settings.
- **Torque:** Transmissions with high torque levels are feasible.

- **High Efficiency:** Due to their low rolling power, planetary gear systems are approximately 95% more efficient.

Disadvantages

- **Noisy Operation:** Planetary gearheads frequently produce more noise when operating. However, other models, like those in the Kavitsu Transmission product line, promise to be quieter.
- **Load Sharing:** Since the gears must always be beveled with helical teeth, load sharing won't work if the gearing is inaccurate.
- **Lubrication:** Grease lubrication is typically required for planetary gears in order to prevent wear. An oil bath option might be preferred by many consumers.
- **Cost:** The price may be increased in order to guarantee low backlash and lifespan. This is typically due to the fact that they are made up of more parts than other kinds of gearboxes.
- **Loads:** High-bearing loads can cause the dead stud to deteriorate quickly.

WORKING PRINCIPLE OF GEAR CHANGE AND GEAR SHIFTING

In a gearbox arrangement, the gear-changing mechanism is crucial for the selection of the correct gear. This system typically involves the following key components:

- **Gear Changing Lever:** The lever that the operator uses to select the desired gear. It is usually connected to the gear-changing mechanism and allows the operator to shift between different gears.
- **Gear Changing Spindles:** These spindles are responsible for transmitting the motion from the gear-changing lever to the gear forks, allowing for the engagement of different gears.
- **Gear Forks:** The gear forks are used to slide the gears along the main shaft to engage with the corresponding gear on the countershaft. They are typically operated by the spindles to ensure smooth gear changes.
- **Gear Housing:** The housing is the casing that contains all the components of the gearbox, including the gears, spindles, and forks. It provides protection and ensures that the components operate in a contained environment.

These components work together to allow for smooth shifting between gears in a gearbox, ensuring that the tractor or vehicle can adapt to different speeds and loads. The gear-changing mechanism is designed to be robust and reliable to handle the stresses of various applications, whether in a tractor or other machinery.

During gear selection, either the gear (in a sliding mesh gearbox) or the sliding dog collar (in constant mesh and synchromesh gearboxes) is moved using the gear-changing lever to achieve the desired speed ratio. The operator controls the gear lever, which features a swivel pin at its lower end. This pin moves within a box-shaped cavity in the selector rod jaws. By moving the gear lever, the swivel pin activates one of the selector rods from the 3-4 rods housed in the gear selector cover.

Each selector rod is equipped with a gear selector fork, which is positioned on the collar-shaped gear or sliding dog clutch, depending on the system. When the selector rod is engaged, the fork pushes or pulls the gear collar to achieve the necessary meshing.

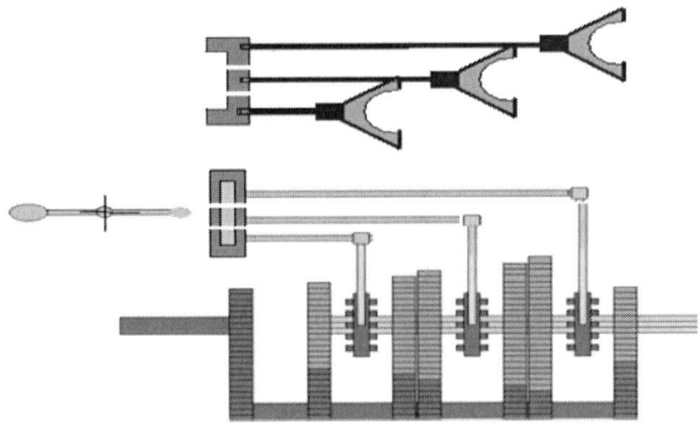

Figure: 10.11 Arrangement Showing the Different Components Involved in Gear Changing Mechanism

Interlock Mechanism

Prevents two gears engaging simultaneously; if this occurs the gearbox will lock up and shaft rotation will be impossible. Although the interlock device takes a number of different forms, the arrangement shown in the figure is one of the most common.

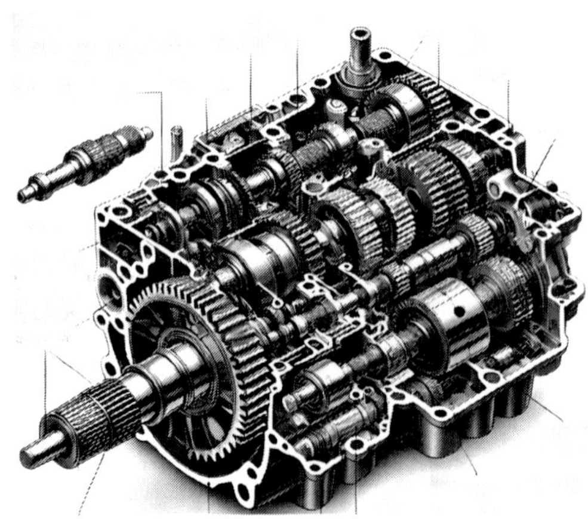

Figure: 10.12 Interlock Mechanism

Calculation of speed ratio

Calculation of speed ratio shows a typical constant mesh gear box in an agricultural tractor. This gear box has 8+2 gear ratios i.e. 8 forward speed gear ratios and 2 reverse speed gear ratios.

Gear Meshing

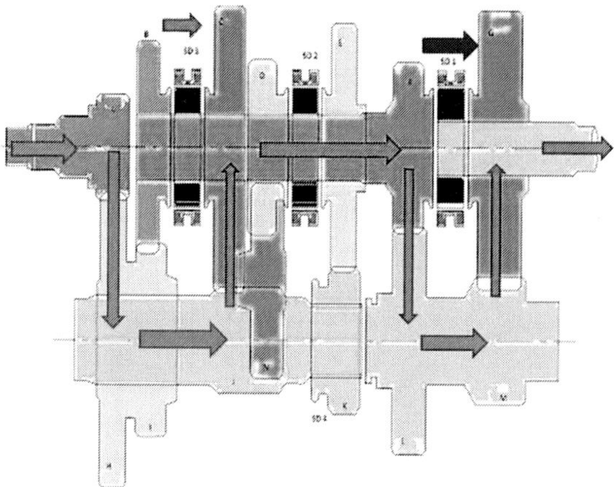

Low Gear	High Gear
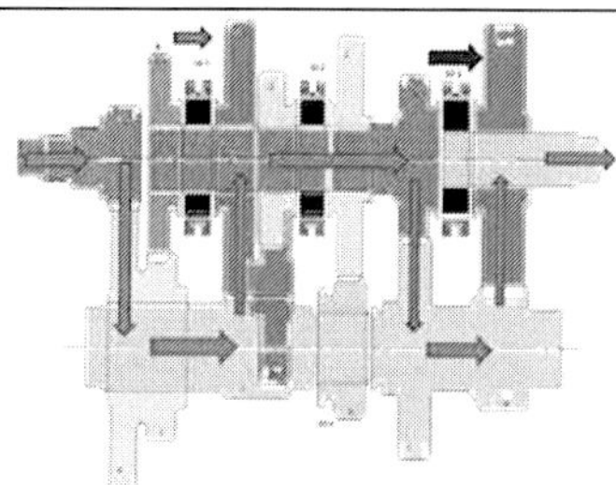	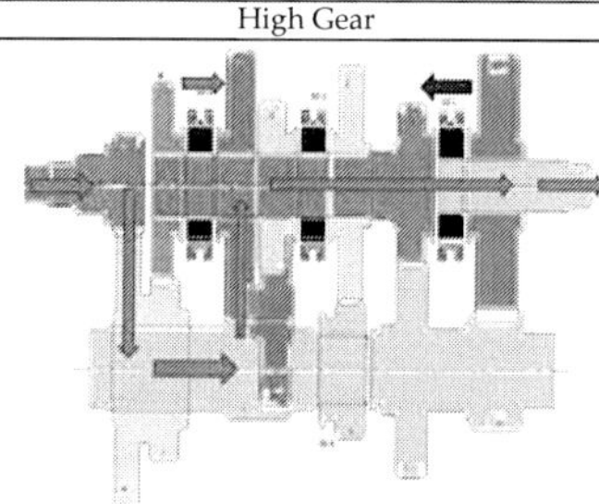
1st Gear	1st Gear
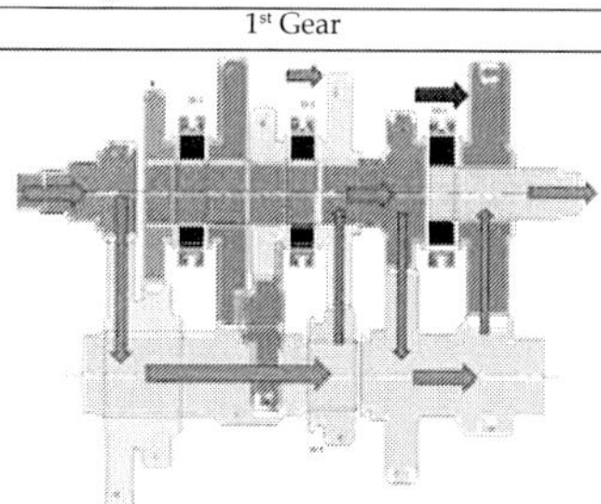	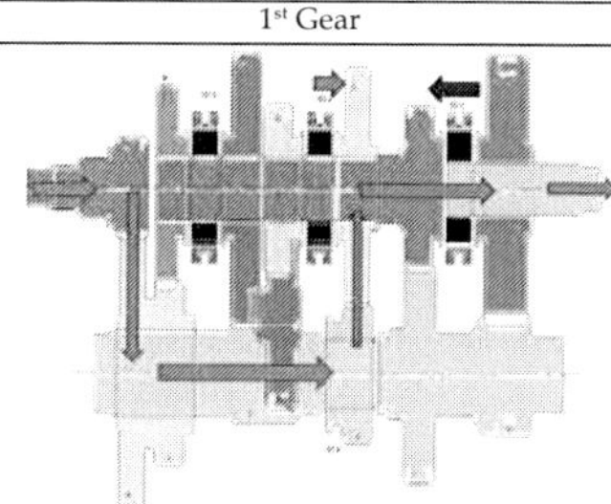
2st Gear	2st Gear
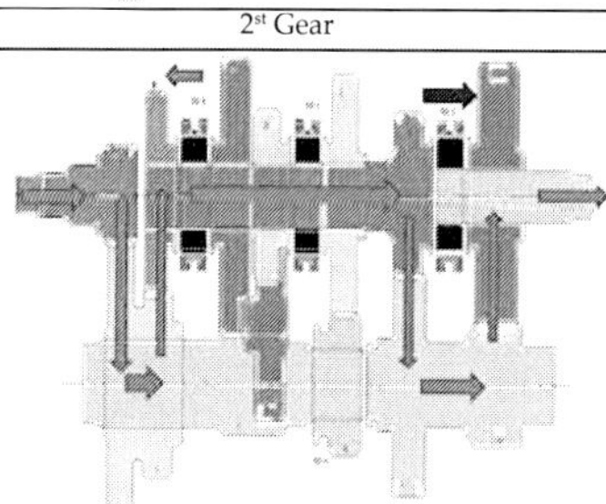	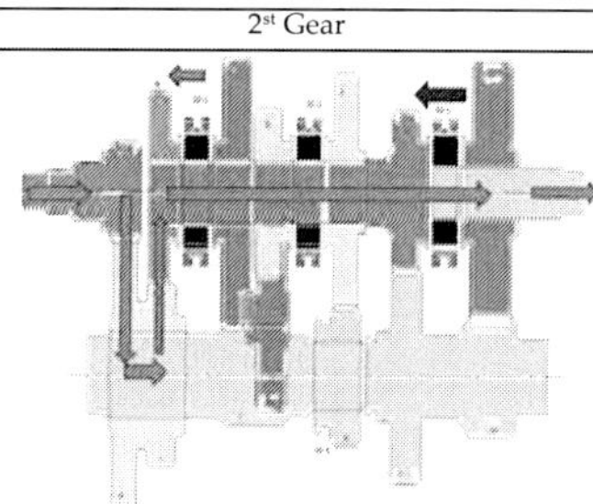

Figure: 10.13 Low Gear , High Gear ,1st Gear , 2nd Gear,

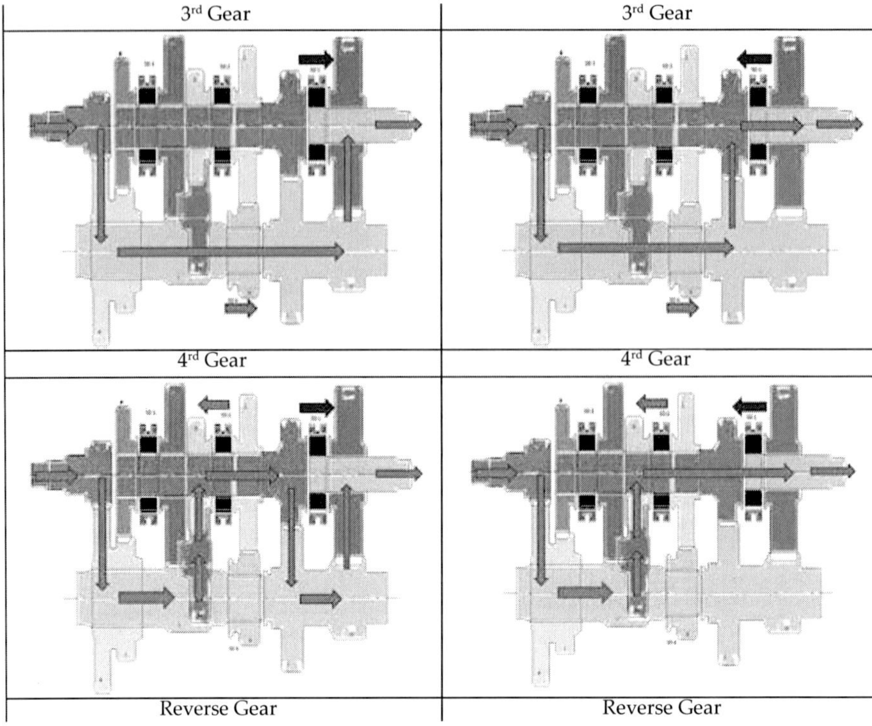

Figure: 10.14 3rd Gear , 4th Gear , Reverse Gear

POWER TRANSMISSION IN POWER TILLER

For operation of power tiller, the power is obtained from the I.C. engine, fitted on the power tiller. The engine power goes to the main clutch with the help of belt or chain. From main clutch, the power is divided in two routes, one goes to transmission gears, steering clutch and then to the wheel. The other component goes to the tilling clutch and then to the tilling attachment.

The flow diagram for transmission of power is given below:

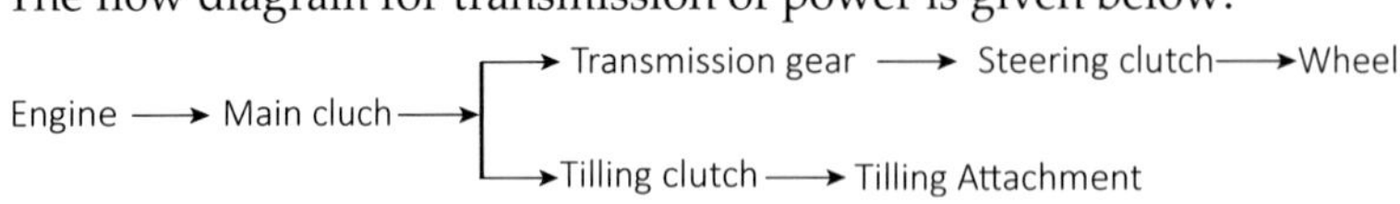

V-belt is usually used to transmit power from the engine to the main clutch, because V-belt has very high efficiency and it works as a shock absorber also.

Main Clutch

Power goes from the engine to the main clutch.

Clutch can be of following two types:

- Friction clutch or

- V-belt tension clutch

Friction clutch is generally used for bigger power tiller. Usually it is a dry type multiple disc clutch. V-belt tension clutch is used for small power tillers.

The main functions of clutch in a power tiller are:

- To transmit engine power to transmission gears and
- To make power transmission gradual and smooth

A. Trouble – Clutch Slips

Some of the causes and remedies are as follows:

1. **Cause:** Oil inside the clutch. Remedy – Clean with gasoline and dry it
2. **Cause:** Worn lining. Remedy – Replace the lining
3. **Cause:** Spring tension not sufficient. Remedy – Replace the spring
4. **Cause:** Water inside the clutch. Remedy – Dry it with heater etc.
5. **Cause:** Poor adjustment of clutch. Remedy – Adjust the clutch
6. **Cause:** Poor contact of friction surface. Remedy – Grind the surface

Trouble – The Clutch is Not Disengaged

Some of the causes and remedies are as follows:

1. **Cause:** Poor adjustment of clutch. Remedy – Adjust it correctly
2. **Cause:** Corrosion or rust on the pressure plate. Remedy – Reface with sand paper or a file
3. **Cause:** Poor contact of friction surface. Remedy – Reface with sand paper or a file

Transmission Gears

Transmission box consists of gears, shafts and bearings.

The Speed Change Device May Be:

- Gear type or
- Belt type

Brakes

All power tillers have some braking arrangement for stopping the movement. Most of the power tillers use Inner side expansion type brake.

Wheels

Usually 2 to 4 ply pneumatic tyres are used in power tillers. The pressure of the tyre ranges from 1.1 to 1.4 kg/cm2.

Rotary Unit

Power tiller has a rotary unit for field operation.

Rotary Unit is of Two Types

- Centre drive type (transmission at the centre)

- Side drive type (transmission at one side)

Centre Drive Type

It has the following characteristics:

1. Tilling width can be widened
2. Rotary unit is light in weight
3. Fixing of attachment is easy
4. The tine shaft can be detached easily
5. Mounting and dismounting of rotary unit is very easy
6. It may leave some portion of the field unfilled
7. It has one point support on the ground.

Side Drive Type

Characteristics of Side Drive Type:

- Deeper tilling is possible
- The arrangement is useful for hard soil
- It has two points support on the ground

Rotary Tines

Rotary tines are used in rotary unit for soil cutting and pulverisation purpose.

Rotary Tines are of Three Types:

- Straight Tines:

  In Case of Straight Tines:
  - Power consumption is less
  - Fine pulverization of soil is possible
  - Poor soil turning
  - Grass entangles in the tines very easily
  - It is suitable for hard soil
- Curved Tines:

  In Case of Curved Tines:
  - Good soil turning is possible
  - It is suitable for avoiding grasses
  - Pulverization of soil is coarse and
  - Power consumption is high

Sliding Tines

Sliding tines have the characteristics of sliding on their positions according to the requirement.

The tines can be arranged in three ways:

- Inner heap (to break the ridges)
- Outer heap (to make the ridges), and

- Even arrangement

Steering Clutch Lever

Steering clutch is provided on the grip of the right and left handles. When the left side is gripped, power is cut-off on left side of the wheel and the power tiller turns to the left. Similarly when the right side is gripped, the power tiller turns to the right.

POWER TILLER

A prime mover, known as a hand tractor or walking-type tractor, is controlled by the operator walking behind it, allowing for field operations. The concept of the power tiller emerged in 1920, with Japan being the first country to adopt it on a large scale. The first successful power tiller model was developed in Japan in 1947. In India, the power tiller was introduced in 1963, and manufacturing of various models, including Iseki, Sato, Krishi, Kubota, Yanmar, and Mitsubishi, began after 1962. Agricultural power tillers are used for tasks such as ploughing, sowing, spraying, harvesting, and transportation, and are particularly essential for puddling in rice cultivation.

Figure: 10.15 Power Tiller

Components of a Power Tiller

1. Engine
2. Transmission Gears
3. Clutch
4. Brakes
5. Rotary Unit

Power tillers are equipped with internal combustion (I.C.) engines. Most modern models, such as those by Kubota, Mitsubishi, Krishi, Yanmar, and Satoh, utilize diesel engines. However, Iseki models are notable for employing kerosene engines.

Operation of a Power tiller (Power Transmission in a Power Tiller)

The power for operating the power tiller is generated by the internal combustion (I.C.) engine installed on it. The engine's power is transmitted to the main clutch via a belt or chain. From the main clutch, the power is split into two paths: one route leads to the transmission gears, steering clutch, and wheels, while the other directs power to the tilling clutch and subsequently to the tilling attachment. A V-belt pulley is typically used to transfer power from the engine to the main clutch.

- **Main Clutch:** Power goes from engine to main clutch through V- belt – pulley arrangement
- **Transmission Gears:** Transmission gears consists of gears, shafts, and bearings. Transmission gears reduce speed of the engine and increase the torque at the wheels.
- **Brakes:** All power tillers have braking arrangement for stopping the movement of power tiller.
- **Wheels:** Usually two to four ply pneumatic tyre are used in power tillers for the wheels. The inflation pressure ranges from 1.1 to 1.4 kg/cm2.
- **Rotary Unit:** Power tillers have a rotary unit for field operation. Rotary tines are used in rotary unit for soil cutting and pulverizing purposes. Rotary tines are of three types
    - Straight tines
    - Curved tines
    - L shape blades

Uses

1. For puddling operation in paddy fields- using rotary tines
2. For cutting and pulverizing the soil in dry lands and in garden lands
3. For cutting and pulverizing the stubbles of sugarcane, maize and cotton
4. For sowing and inter-cultivation works
5. For spraying of orchard trees
6. For transporting purposes

11

Power Take Off Drive

INTRODUCTION

A PTO (Power Take Off) in a tractor transfers mechanical power from the engine to various tractor implements or attachments, such as a rotavator, thresher, straw reaper, or super seeder, through a PTO shaft. In India, the PTO is typically located at the rear of the tractor, while in several other countries, a front PTO is also available to power front-mounted machines. Different PTO-driven implements require varying amounts of PTO horsepower (HP) for operation. PTO horsepower indicates the power available for driving tractor attachments, and it is transmitted to these implements via the PTO shaft.

The rear-mounted power outlet, known as the PTO, is used to operate PTO-driven implements like a rotavator, power harrow, and baler. Tractors are used not only for towing implements for tillage and sowing but also for stationary tasks such as crop threshing and water pumping from tube wells, which are powered by the PTO. The PTO-driven implements require a specific amount of horsepower, referred to as PTO HP, to function. This PTO HP is the portion of the engine power used to operate the implements, and it is always lower than the tractor's total horsepower due to friction losses. Additionally, the PTO shaft derives its power from the gearbox.

There are two standard PTO speeds: 540 rpm and 1000 rpm. This standardization was introduced to ensure that implements are designed to operate efficiently at these speeds. For splines, the 540 rpm shaft has 6 splines, while the 1000 rpm shaft features 21 splines.

THE INVENTION OF PTO IN THE TRACTOR INDUSTRY

In 1918, the International Harvester Company (IHC) introduced the Power Take Off (PTO) in tractors, with Edward A. Johnston, an IHC engineer, being the first to develop the concept after being inspired by homemade PTO systems in France. The 8-16 model became the first commercial tractor to feature PTO. Two years later, in 1920, the company released the 15-30 model, which was the first PTO-equipped tractor to undergo the Nebraska tractor test.

The need for standardization arose so that equipment could be designed to operate at specific speeds. Initially, the standard PTO speed was 536 RPM, but in 1958, it was changed to 540 RPM, and a 1000 RPM PTO speed was introduced as well. Over time, other tractor manufacturers adopted the PTO system, making it an integral component in tractors worldwide.

How does a PTO Work?

As we understood, a tractor PTO provides the mechanical power of the tractor engine to the implement or attachments. Now, let's understand how this energy transfers from the engine to various implements. So, the power is transferred by joining the driveline shaft (connecting rod) of the implement to the PTO shaft of the tractor. When the tractor engine starts, the driveline shaft rotates, giving a rotational force to the implement attached through the Universal or 'U' joint to the PTO shaft, and instantly, the attached implement starts working. The most important point is that only those implements require PTO that need rotational force to operate.

Power Take off

The Power Take Off (PTO) is an essential component of a tractor's transmission system. It consists of a shaft, a shield, and a cover. The shaft is externally splined to transmit torsional power to another machine. To ensure safety, a rigid guard called the power take-off shield is fitted around the PTO shaft to protect operators. According to ASAE standards, the PTO speed should be 540 ± 10 rpm when operating under load. Additionally, a new standard has been established for operating a 1000 rpm PTO-driven machine.

Figure: 11.1 PTO on a Tractor

Figure: 11.2 PTO Drive

Belt Pulley

All tractors are equipped with a belt pulley, which serves the purpose of transmitting power from the tractor to stationary machinery through a belt. It is commonly used to operate equipment such as threshers, centrifugal pumps, silage cutters, and various other machines. The pulley can be located on the left, right, or rear side of the tractor. To engage or disengage the pulley drive from the engine, a clutch mechanism is used.

Figure: 11.3 Belt Pulley Connected to Tractor PTO

Control Board or Dash Board

The control board of a tractor generally consists of

1. Main Switch

When the main switch is on , the electric current floes in the electric circuit

2. Throttle

This lever is for increasing or decreasing the speed of the engine.

3. Decompression Lever

This lever releases the compression pressure from the combustion chamber of the engine and helps to start the engine.

4. Hour meter

This meter indicates the engine hour as well as engine revolution per minute

5. Light Switch

Light switch is for light points only.

6. Horn Button

This is for horn of the tractor.

7. Battery Charging Indicator

This indicates the charge and discharge of the battery.

8. Oil Pressure Indicator

This indicates the lubricating oil pressure in the system.

9. Water Temperature Gauge

Gauge: this indicates the temperature of water of the cooling system.

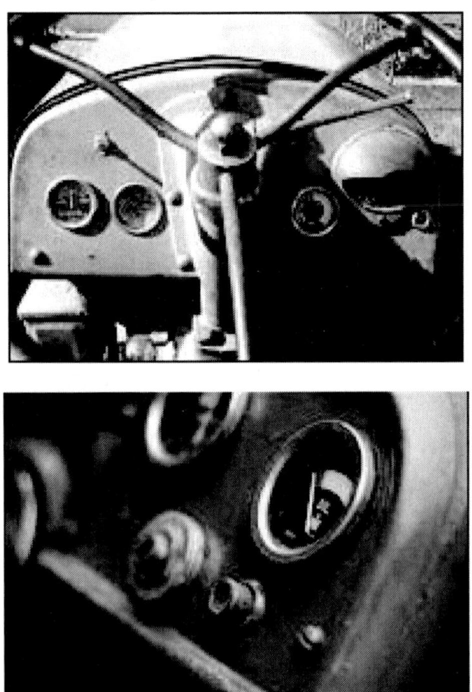

Figure: 11.4 Tractor Dash Board

IMPORTANT TERMS CONNECTED WITH TRACTORS

- **Wheel Base:** The wheel base refers to the horizontal distance between the front and rearwheels of a tractor.

- **Turning Space:** Turning space is the diameter of the smallest circle traced by the outermost point of the tractor when moving at a speed not exceeding 3 km/h with the steering wheels fully locked.
- **Cage Wheels:** Cage wheels are wheels or attachments to wheels that feature spaced crossbars, designed to improve the traction of the tractor in wet fields, and are commonly used in paddy fields.
- **Ground Clearance:** Ground clearance is the height from the lowest point of the tractor to the ground surface, when the tractor is loaded to its maximum permissible weight.
- **Track:** Track is the distance between the two wheels on the same axle of the tractor, measured at the point where the wheels make contact with the ground.

CRAWLER TRACTOR

A crawler tractor, also known as a track-type tractor, tractor crawler, or track-laying vehicle, operates on continuous tracks rather than wheels. In agriculture, it is primarily used for land clearing and leveling. The key advantage of crawler tractors over wheeled tractors is that they have a larger surface area in contact with the ground, which results in a significantly lower force per unit area compared to conventional wheeled tractors of the same weight. This makes crawler tractors ideal for soft, low-friction, and uneven terrains like mud, ice, and snow. However, the main drawback of crawler tractors is that their track system is more complex than wheels and is more prone to failure issues such as snapped or derailed tracks.

Characteristics

Crawler tractor or chain type tractor has the following characterizes:

1. It is most suited for heavy work, specially earth moving work and reclamation work
2. It is used for all types of agricultural works with heavy implements
3. It is designed to secure good adhesion and transmit high drawbar pull in difficult field conditions, where wheel tractors fail to secure adequate grip on the soil
4. It provides large area of contact with the ground
5. It is useful at places where adhesion is difficult and rolling resistance is high

VARIOUS PTO TYPES IN A TRACTOR

The PTO is divided into two types, namely, Live PTO and Independent PTO. Let's understand each of the types below.

Live PTO

The live PTO has two types of transmission clutch. If it is pressed halfway, the transmission will stop, and PTO continues spinning under a load. When pressing the clutch fully, the PTO also stops working. Simply, a live PTO separates the transmission's control from the PTO operation control.

Independent PTO

The other PTO type is an Independent PTO or IPTO. In this type of PTO, farmers can work independently with the tractor and the implement. They can run the PTO even if the tractor is not moving. This PTO can be engaged with the help of a button or lever that controls the power take-off. IPTOs are a great option for continuous operations as in this PTO type, the PTO is not affected by the position of gears, and a driver can switch gears when running PTO. The Independent PTO has two types: mechanical and hydraulic.

WHAT ARE THE DIFFERENT PTO SPEEDS?

There are different PTO speeds required for running a tractor. Let's have a look at each of the speeds.

Standard PTO

The standard PTO speed of 540 RPM refers to a system where both the PTO of a tractor and the implement driveline shaft rotate at 540 RPM at the specified maximum engine RPM (ERPM). This speed is optimal for everyday agricultural tasks like ploughing. To operate at this speed, a 6-spline connection is required for the PTO shaft.

Economy PTO

This is the PTO speed of 540 RPM on a lower engine RPM when compared to the standard PTO speed. For example, for a PTO speed of 540 RPM, in the standard PTO it will be achieved at 2100 engine RPM whereas in case of economy PTO the same speed will be achieved at 1600 RPM. This results in lesser fuel consumption. This PTO speed is best for applications like puddling, generators and water pumps.

Reverse PTO

Reverse PTO is helpful in clearing any obstruction that is stuck with the implements while operating. It is used through a lever or button given in the tractor.

MSPTO and GSPTO

The MSPTO stands for Multi-Speed Reverse PTO, which gives a driver various forward and reverse gear speeds for power take-off. It made the farming process easy and more productive. Whereas GSPTO stands for Ground Speed PTO, in this case the PTO speed is same as that of the tractor's ground speed.

12

Braking System

INTRODUCTION

The brake is a crucial component used to stop or slow down the movement of a tractor. It is typically mounted on the driving axle and is controlled by two independent pedals. Each pedal can be operated separately to help with turning the tractor during field operations, or they can be locked together using a locking mechanism. The two main types of brakes are: a) Mechanical brake, and b) Hydraulic brake.

Figure: 12.1 The braking system plays a critical role in the transmission system of a tractor, as it is responsible for slowing down or halting the tractor's movement. Brakes function by converting the kinetic energy of the moving tractor into heat through friction between a rotating and a stationary component. Let's explore the commonly used braking systems in tractors.

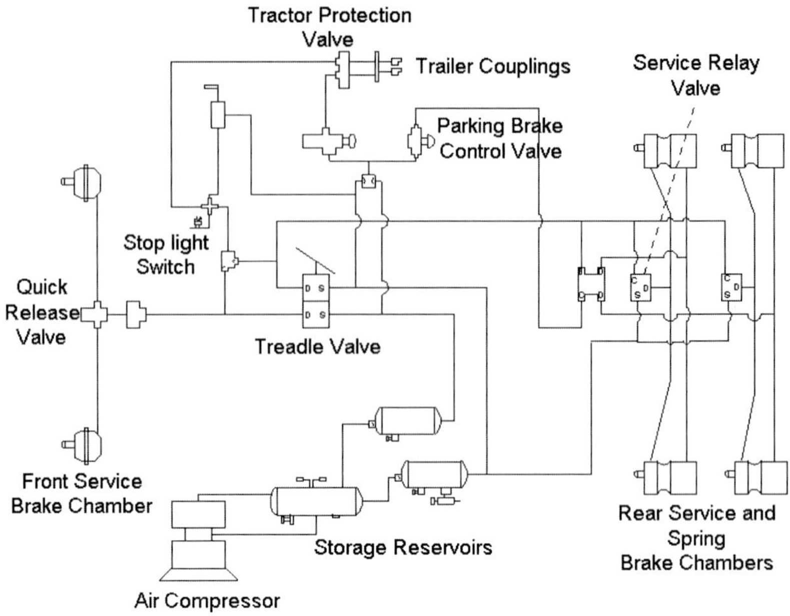

Figure: 12.2 A general Layout of a Tractor Air Brake System

THE IMPORTANCE OF TRACTOR BRAKE SYSTEMS

Tractors are frequently used in expansive fields, which means they often cover long distances and encounter challenging terrain. In these situations, the tractor brake system becomes essential. Equipped with powerful engines that can run at high speeds, tractors require an efficient braking system to maintain control. Without it, the risk of losing control increases significantly. Tractor brake systems ensure that farmers can operate and maneuver their vehicles safely, enabling them to carry out agricultural tasks more effectively and with greater safety.

Principle of Operation

Brake works on the principle of friction. When a moving clement is brought into contact with a stationary element, the motion of the moving element is affected. This is due to frictional force, which acts in opposite direction of the motion and converts the kinetic energy into heat energy.

Classification of Brake

Brake can be classified as:

1. Mechanical brake.
2. Hydraulic brake.

Mechanical Brakes

Mechanical brakes function by utilizing levers or linkages. When the operator presses the brake pedal, it activates the linkages, which then push the brake shoes against the brake drum, creating friction that slows down or halts the tractor. There are three types of mechanical brakes: internal expanding shoe type, external contracting shoe type, and disc type. However, the internal and external shoe types are now considered outdated technologies, with disc brakes being the more common choice. While mechanical brakes are cost-effective and straightforward, they require regular maintenance and adjustments.

Hydraulic Brakes

The hydraulic brake system is based on Pascal's law principle. It uses fluid pressure to operate. The brake fluid is basically a mixture of alcohol and glycerin which is filled in the master cylinder. The hydraulic fluid is forced into the cylinder at each wheel when the brake pedal is pushed. It then pushes the brake pads against the discs or drums. The hydraulic brakes are quite reliable and need less manual adjustment than mechanical brakes. They are also better at heat dissipation. However, they are complex, expensive, and high maintenance.

Mechanical brake can be:

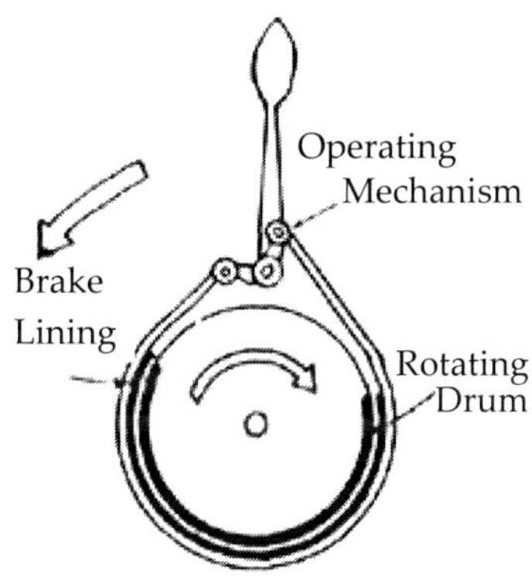

Figure: 12.3 (a) Internal Expanding Shoe type ,(b) External Contracting Shoe type and ,(c) Disc Type.

The brakes used on automobiles can be classified in a number of ways. Some of these classifications are mentioned here under:

Purpose

Brakes can be classified on the basis of the purpose for which they are on the automobile:

- Primary Brakes: These are used as the main brakes to stop the motion of the automobile.
- Secondary Brakes: These are used as parking brakes to maintain the status of the automobile specially when parked on slopes or on even surfaces.

Location

The brakes can be located at the wheels or on the drive shaft of the transmission. The location of brakes on the transmission shaft not only gives the convenience of location and provides for equal braking effort on the braking wheels but locating these on the wheels provides dual advantage of having higher surface area for brake application and the greater area provides for better heat dissipation.

Construction

The brakes are classified into:

Internal Expending Shoe Brakes

Two brake shoes made of frictional material fitted on the inside of the brake drum are held away from the drum by means of springs. One end of each shoe is fulcrum whereas the other is free to move by the action of a cam which in turn applies force on the shoes. The movement of the cam is caused by the brake pedal through the linkage. The drum is mounted on the rear axle whereas the shoe assembly is stationary and mounted on the back plate.

External Contracting Shoe Brakes

This type of brake system is normally available on crawler tractors. The brake band directly surrounds the drum mounted on the drive axle. When the pedal is depressed, the band tightens the drum

Disc Brakes

The internal expanding shoe brakes, also known as drum brakes, are the most commonly used type in tractors. However, newer tractors are being equipped with disc brakes. In disc brakes, two actuating discs have holes drilled into them, with steel balls placed in these holes. When the brake pedal is pressed, the links cause the two discs to move in opposite directions, which pushes the steel balls into the shallow sections of the holes. This action expands the discs, and braking discs are then pressed between the two expanded discs and the stationary housing. The braking discs are directly mounted on the differential shaft, transferring the braking effect to the differential shaft. On the other hand, external contracting shoe brakes, which are used in railways, are located outside the wheel.

Method of Actuation

The method of actuation decides as to how the braking effort from the point of application brake pedal, reaches tree brakes.

3. **Mechanical Actuation:** Most common in tractors.
4. **Hydraulic Actuation:** used in tractors with higher HP and motor cars.

Hydraulic brake system is based on the principle of pascal's law. The brake fluid, which is usually a mixture of glycerin and alcohol, is filled in the master cylinder. When the pedal is depressed, the piston of the master cylinder is forced into the cylinder and the entire system turns to a pressure system. Immediately, the piston of the wheel cylinder slides outward which moves the brake shoes to stop the rotating drum. When the pedal is released, the return spring of the master cylinder moves the piston back to its

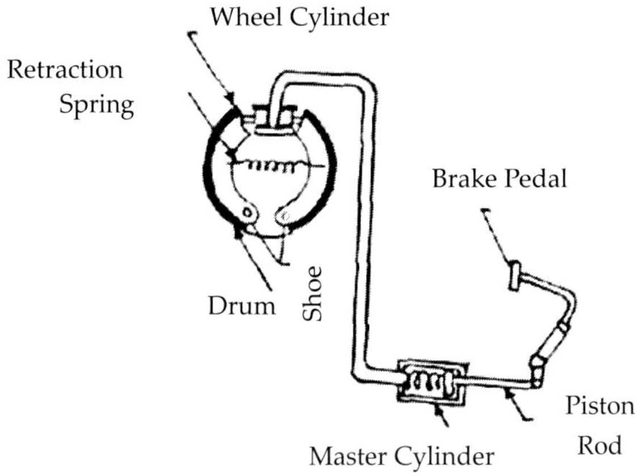

Figure: 12.4 Master Cylinder

- Electric Actuation: use eddy currents for application of brakes.
- Vacuum Actuation: applied in railways.
- Pneumatic Actuation: applied in heavy vehicles and special purpose machines.

Tractor brake systems are generally categorized into two main types: hydraulic and mechanical.

- Hydraulic Brake Systems: These systems use hydraulic oil to perform braking and are commonly preferred in disc brakes.
- Mechanical Brake Systems: These systems use mechanical connections for braking and are typically found in drum brake systems.

Both systems require periodic maintenance and inspections to operate effectively.

Regular inspections of tractor brake systems ensure the early detection of potential issues. Elements such as the wear status of brake pads, brake fluid levels, and deformation of brake discs should be checked during routine maintenance. Adhering to the periodic maintenance schedules recommended by the manufacturer is crucial. This ensures the tractor's longevity and safe operation.

Braking Efficiency

High braking efficiency is essential, especially in emergency situations where quick stopping is required. However, while increased braking efficiency results in shorter stopping times, it can also lead to injuries for the driver due to high decelerating forces and cause loads in the trolley to become dislodged. Additionally, higher braking efficiency accelerates brake wear and increases the risk of losing control of the vehicle. Braking efficiencies in the range of 50-80% are typically sufficient for stopping within a reasonable distance. The actual stopping distance, however, is influenced by road conditions and tire condition.

Braking distance refers to the distance a vehicle travels from the moment the brakes are fully applied to when it comes to a complete stop. It is primarily influenced by the vehicle's initial speed and the friction coefficient between the tires and the road surface. Braking distance also takes into account the reaction time—when the driver perceives the need to stop—and the response time.

HOW TO ENHANCE BRAKE SYSTEM SAFETY

To increase the safety of brake systems, attention should be paid to the following topics:

Safety Enhancement with Innovative Technologies

Tractor manufacturers today use technology to improve tractor brake systems further. Innovative features such as electronic brake control systems, anti-skid technologies, and automatic braking systems enhance the safety standards of tractors, providing operators with more control.

User-Friendly Designs of Brake Systems

User-friendly designs of tractor brake systems make it easier for operators to effectively use these complex systems. Modern tractors come equipped with ergonomic brake pedal arrangements, responsive brake systems, and user interfaces. These features provide operators with both a safe and comfortable driving experience.

Brake System Training and Awareness

Training programs on brake systems for farmers ensure that operators understand and use tractor brake systems correctly. Having the correct information about brake systems helps operators identify potential hazards in advance and prevent them. These training programs also assist operators in performing maintenance routines correctly and detecting potential issues at an early stage.

Legal Standards and Safety Regulations

Tractor brake systems must comply with various national and international safety standards. Manufacturers must adhere to the standards set when designing and producing tractors. Additionally, safety regulations and standards for operators specify the rules to be followed during tractor use. Adhering to these standards is a fundamental element in ensuring the safety of farmers and other workers.

Environmental Impacts and Emission Control

Tractor brake systems should consider environmental impacts as well. An effective brake system helps control the tractor's emission levels, contributing to environmentally friendly use. Manufacturers should improve tractor brake systems by using eco-friendly materials and increasing energy efficiency.

In farming, the safe use of tractors is crucial for both operational efficiency and operator health. Regular maintenance and inspection of tractor brake systems are essential to ensure the safety of farmers and the prolonged use of tractors. The integration of innovative technologies into tractor brake systems provides farmers with increased safety and control, contributing to the sustainable and efficient development of the agricultural sector.

13

Steering System

INTRODUCTION

The system responsible for controlling the angular movement of the front wheels of a tractor is called the steering system. This system reduces the effort required by the operator to turn the front wheels through the application of leverages. The key components of the steering system include the steering wheel, steering shaft, steering gear, drag link, steering arm, tie rod, and king pin.

When the operator turns the steering wheel, the motion is transferred via the steering shaft to produce angular movement in the pitman arm, through a set of gears. The pitman arm then transmits this angular motion to the steering arm via the drag link and tie rods. The steering arm, which is attached to the king pin (integrated into the stub axle where the wheels are mounted), influences the movement of the front wheels.

The primary function of the steering system is to control the vehicle's direction. This is achieved by converting the rotation of the steering wheel into a change in the angle of the steering wheels' axes through a series of links. The steering system also ensures directional stability, ensuring that the vehicle automatically returns to a straight path when the steering force is released. Additionally, the system is designed to minimize tire wear.

Steering systems can be classified based on their wheel operation: front wheel steering, rear wheel steering, or all-wheel steering.

When the operator turns the steering wheel, the motion is transmitted through the steering shaft to tire angular motion of the pitman arm, through a set of gears. The angular movement of the pitman arm is further transmitted to the steering arm through the drag link and tie rods. Steering arms are keyed to the respective king pins which are integral part of the stub axle on which wheels are mounted. The movement of the steering arm affects the angular movement of the front wheel. In another design, instead of one pitman arm and drag link, two pitman arms and drag links are used and the use of tie rod is avoided to connect both steering arms.

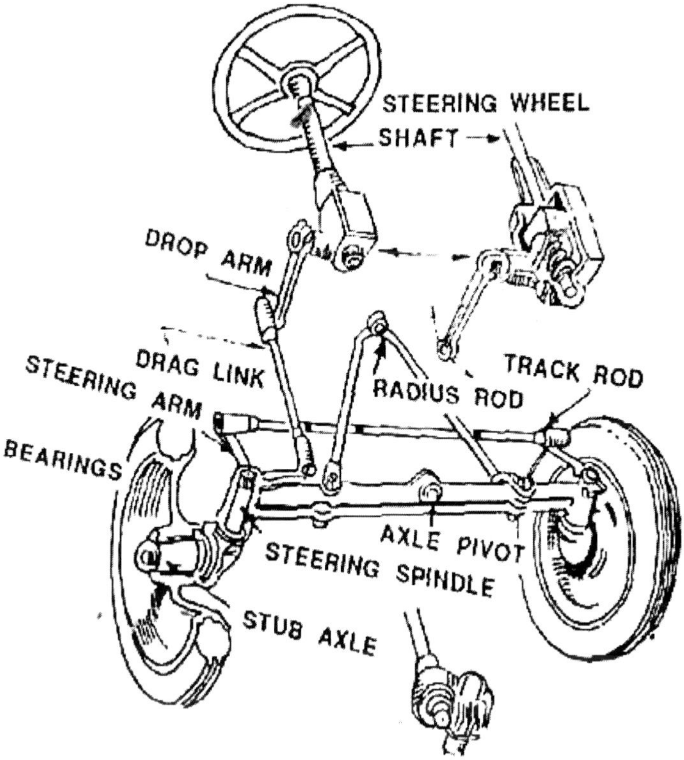

Figure: 13.1 Steering System

- Steering wheel
- Steering shaft
- Steering gear
- Pitman arm (drop arm)
- Drag link
- Steering arm
- Tie rod and
- King pin.

FUNCTIONS OF A TRACTOR STEERING SYSTEM

- **Control the direction of motion:** The major function of the steering system is to control the direction of motion of the tractor.
- **Provide Directional Stability:** The steering system is also responsible for providing directional stability. The motion of the vehicle being steered needs to become straight ahead when the force on the steering wheel is removed.

- **Improve Handling:** The steering system helps in controlling the vehicle. Improper handling can be due to vehicle pulling on one side, oversteering, or understeering.
- **Improves Tire Life and Performance:** Proper rolling tire contact on the road, as well as the prevention of tire slippage due to faulty alignment, can increase tire life and performance.

Components of a Tractor Steering System

The tractor steering system is a critical element that plays a central role in controlling the angular motion of the front wheels, allowing for precise navigation and manoeuvrability in agricultural operations.

Its primary function is to convert the rotary motion of the steering wheel into the directional movement of the tractor, thereby minimizing the operator's effort in steering and handling the machine.

Key Components of the Tractor Steering System

Steering Wheel

The steering wheel serves as the primary interface between the operator and the tractor's steering system. It is the point where the operator applies rotational force to initiate the steering process.

Steering Gear

The steering gear, often located near the steering wheel, is responsible for converting the rotational motion into lateral movement. It plays a crucial role in determining the responsiveness and precision of the steering system.

Steering Shaft

The steering shaft acts as a conduit, transmitting the rotational input from the steering wheel to the steering gear. It ensures a seamless transfer of motion, allowing for a direct connection between the operator and the steering mechanism.

Drop Arm (Pitman Arm)

The drop arm, also known as the Pitman arm, is a pivotal component that connects the steering gear to the steering linkage. It converts the rotary motion from the steering gear into the linear motion needed to control the front wheels.

Steering Arm

The steering arm connects the steering linkage to the front wheels. It plays a crucial role in distributing the steering force to the wheels, facilitating smooth and controlled turns.

Drag Link

The drag link is a connecting rod that transfers motion between the drop arm and the steering arm. It ensures synchronized movement, allowing for coordinated turning of the front wheels.

King Pin

The king pin is a pivot point around which the front wheels rotate. It is a crucial part of the steering system which enables the wheels to turn smoothly and efficiently.

Tie Rod

The tie rod is a crucial linkage component that connects the steering arms on both sides. It ensures that both front wheels turn together, maintaining alignment and stability during steering manoeuvres.

TRACTOR TYRE AND FRONT AXLE

Tractor tyres come in various sizes, with ply ratings of 4, 6, or 8. The ply rating reflects the relative strength of the tyres, with higher ratings indicating stronger tyres. For example, a tyre size of 12-38 means the tyre has a sectional diameter of 12 inches and is mounted on a 38-inch diameter rim. The inflation pressure for the rear tyres typically ranges from 0.8 to 1.5 kg/cm^2, while for the front tyres, it ranges from 1.5 to 2.5 kg/cm^2. Under normal operating conditions, pneumatic tyres usually last around 6,000 working hours when used for drawbar work.

FRONT AXLE AND STEERING MECHANISM

Front Axle

Front axles are typically dead axles The front wheel hubs rotate on anti-friction bearings of the tapered roller type, which are mounted on the steering spindle. This spindle is part of the steering knuckle assembly. To allow the wheels to turn via the steering gear, the steering spindle and knuckle are hinged at the axle's end. The pin acting as the pivot for this hinge is referred to as the Kingpin or steering knuckle pin.

FRONT WHEEL ALIGNMENT

Proper front wheel alignment is essential for smooth steering, vehicle stability, and minimizing tyre wear. Front wheel alignment is achieved by adjusting the following factors accurately:

- Caster
- Camber
- Kingpin inclination
- Toe-in
- Toe-out

CASTER

The caster angle refers to the tilt of the kingpin or ball joint's centerline from the vertical, either towards the front (negative caster) or towards the rear (positive caster) of the vehicle. This angle generates a trailing effect, contributing to the vehicle's directional stability. Incorrect caster angles can lead to issues such as stiff steering and pulling to one side when brakes are applied. The caster angle typically ranges from 2-8° negative.

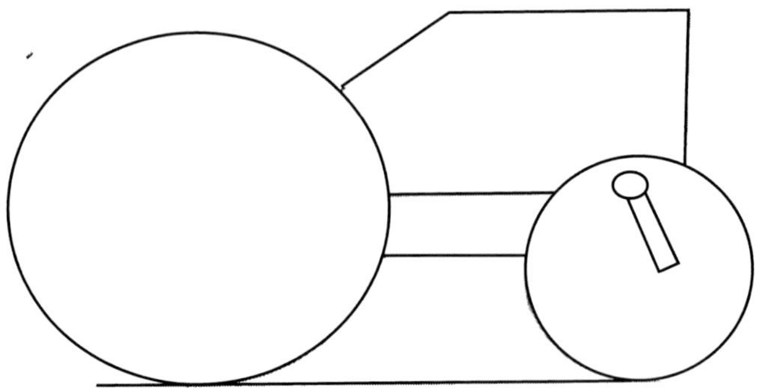

Figure: 13.2 Caster

CAMBER

Camber angle is the inclination between the centre line of the tire and the vertical line. The outward inclination is called positive camber and the inward inclination is the negative camber. The purpose of the camber is to prevent the top of wheels from tilting inward much due to excessive load or play in the king-pin and wheel bearing. Unequal camber in the wheels causes the vehicles to roll in the direction of wheel having the greater camber which upsets directional stability and tends to scuff the tread on the opposite tire, excessive camber prevents the tire from having correct contact with the road which causes it to wear only on the side directly beneath the load.

Camber angle is less than 1 degree.

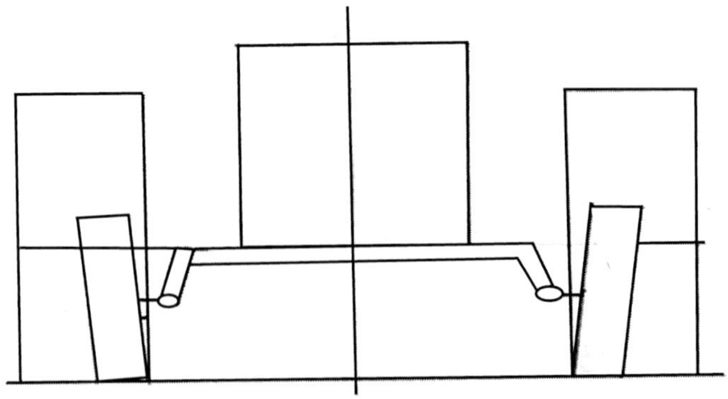

Figure: 13.3 Camber

KING-PIN INCLINATION

King-Pin Inclination refers to the inward tilt of the king-pin or ball joint center line from the vertical axis. This angle, when combined with camber, plays a crucial role in providing directional stability for the vehicle. The combination of king-pin inclination and camber ensures that the tire's steering is centered and the upward thrust on the stub axle is aligned more closely with the center of the king-pin. Together, these angles contribute to a better steering response and stability. The combined angle of king-pin inclination and camber is known as the included angle.

King-pin inclination ranges from 4-8°.

TOE-IN

Front wheels are usually drawn in slightly in front so that the distance between the back-ends (y) in slightly more than the distance between front-ends (x). The different between these distances is called toe in. Wheels are toed-in to offset the tendency for them to roll outward due to camber and to play in the steering linkage. Toe-in is usually 2-4 mm.

TOE-OUT

When a vehicle takes a turn, the inside wheel moves faster than the outer wheel because the former has to negotiate an are with shorter radius than the latter. This action causes the wheels to out-out on turns because of difference in their turning angles.

STEERING SYSTEM

The function of a steering system is to convert the rotary movement of the steering wheel in driver's hand into the angular turn of the front wheels on road. Further, the steering system should provide mechanical advantage over front wheel steering knuckles, offering driver easy turning of front wheels with minimum effort in any desired direction.

MECHANICAL STEERING (ACKERMAN-TYPE)

When driver turns the steering wheel, motion is transmitted down through the steering tube to the steering gear. The steering tube revolves inside the steering column. The steering gear changes the direction of motion and increase the turning force applied by driver of the steering wheel in accordance with gear ratio. The gear rotates the steering arm (pitman arm) which transfers the motion to the steering knuckles through the steering gear connecting rod, tie-rod and knuckle arms. This type of linkage is called the relay steering linkage.

The king-pin torque (T) required to turn the wheel under a vertical load (W) can be calculated by the following equation:

$$T = W.f\left(\frac{I_0}{A} + e^2\right)^{1/2}$$

Where,

f = effective friction coefficient

I0 = polar moment of inertia of tire print, m4

e = Kingpin-offse,t m

A = tire print area, m2

This equation assumes a uniform tire print pressure. If the exact shape of the tire print is unknown, an approximation can be made by assuming it to be circular with diameter equal to the nominal tire width, b, then

$$\frac{I_0}{A} = \frac{\pi}{32}d^4 \times \frac{4}{\pi d^2} = \frac{d^2}{8} = \frac{b^2}{8}$$

So equation (1) becomes

$$T = wf\left(\frac{b^2}{8} + e^2\right)^{1/2}$$

The relationship between steering torque and king-pin offset is depicted in the curve below. This curve indicates that there is an optimal king-pin offset (denoted as e) that results in the minimum steering torque (T). The ideal offset varies depending on the surface conditions. For situations involving heavily loaded tyres on loose soil, where rolling resistance is high, a small king-pin offset or even center point steering (where the offset approaches zero) is more effective. Additionally, if the steered wheel is powered or equipped with brakes, a small or zero offset is recommended to minimize the feedback of forces into the steering system.

The curve below also demonstrates the generally accepted typical effective coefficient of friction for rubber tyres on dry concrete, in relation to the king-pin offset to tyre width ratio.

Importance of Wheel Alignment

- **Improves Handling:** Proper wheel alignment helps in better control of the vehicle. Misalignment can cause the vehicle to pull to one side or cause steering wheel vibrations, both of which affect handling.
- Enhances Tyre Life and Performance: Correct alignment ensures the tyres make proper contact with the road, preventing slippage and excessive wear, ultimately improving their longevity.
- Identifies Potential Problems: Wheel misalignment can often indicate underlying issues with the vehicle. Regular checks can help detect and correct faults before they become more serious.

- **Ensures Safety:** Keeping the alignment in good condition ensures that all vehicle systems, especially the suspension, are functioning properly, thereby improving overall safety.
- **Boosts Fuel Efficiency:** When the vehicle's systems, including wheel alignment, are optimized, fuel efficiency improves, leading to lower operating costs.

TYPES OF TRACTOR STEERING SYSTEMS

Manual Steering

Manual steering systems are the traditional and simplest type found in older tractors. They rely on mechanical linkages such as steering columns, gears, and rods. The operator exerts physical force to turn the steering wheel, directly controlling the movement of the front wheels.

Hydraulic Power Steering

Hydraulic power steering systems use hydraulic pressure to assist in steering. The hydraulic power steering system is a more modern system that is used in most tractors today. It uses hydraulic pressure to assist the operator in turning the steering wheel.

The system consists of a hydraulic pump, hydraulic cylinder, and control valve. When the operator turns the steering wheel, the motion is transmitted through the steering shaft to the control valve.

Electric Power Steering

Electric power steering (EPS) systems have gained popularity due to their efficiency and ease of integration with modern tractors. EPS systems use an electric motor to assist the steering mechanism. They are known for their responsiveness and the ability to adjust steering efforts based on operating conditions.

What is Power Steering in a Tractor?

In tractors, power steering facilitates easier wheel rotation by using a power source. It may be simpler to turn the wheel with power steering, particularly when moving slowly or towing large objects. Now, to help turn the front wheels, power steering can use hydraulics, which is the application of fluid pressure or electric power.

Components of Tractor Power Steering System

Power steering comes attached with a hydraulic power steering, including components: a pump, cylinder, and control valve. It also has an electric power steering which comprises an electric motor, steering column along with a control module. Learn more about the components below:

Hydraulic Power Steering:

- **Hydraulic Pump:** This pump directs fluid to assist with steering.

- **The Hydraulic Cylinder:** It functions by exerting force on the tractor's framework to turn the wheels.
- **Steering Control Valve:** The valve controls the liquid flow to the cylinder.

Electric Power Steering:

- **Electric Motor:** It assists in turning the wheels.
- **Steering Column:** The part of the tractor where you hold onto the steering wheel.
- **Control Module:** This controls how much assistance the electric motor provides.

How Does Power Steering Work?

Power steering in tractors enables easy movement of the front wheels in response to turning the steering wheel. The system works by using either hydraulic or electric components to make steering easier and more precise.

- **Hydraulic Systems:** In this type of system, fluid power is used to assist steering. When the operator turns the wheel, pressurized hydraulic oil is pumped through tubes to a cylinder that contains a piston. The hydraulic oil enhances the force applied to the steering, making it easier to turn the wheels.
- **Electric Systems:** Instead of hydraulic fluid, electric power steering (EPS) uses an electric motor to assist in steering. A computer module controls the motor, which applies assistive torque based on sensor readings that measure steering wheel torque and position. The assistance level adjusts according to driving conditions, and EPS systems tend to offer better fuel efficiency, simplified manufacturing, and the ability to integrate additional driver assistance features.

Comparing the Two Power Steering Systems

- Electric Power Steering (EPS) uses an electric motor and provides more precise steering, smoother operation, and lower maintenance needs. Newer tractors often utilize EPS systems for improved accuracy and efficiency.
- Hydraulic Power Steering (HPS) uses fluid pressure to help with steering. Although older tractors often have this system for powerful assistance, it may lack the precision of newer electric systems.

Benefits of Power Steering in Tractors

- **Effortless Steering:** Power steering reduces operator fatigue by requiring minimal effort to turn the wheel, leading to more efficient and accurate task completion.
- **Improved Maneuverability:** Power steering offers smooth, precise handling, helping the operator navigate around obstacles and through the field with ease.
- **Increased Performance:** By minimizing fatigue and increasing efficiency, power steering allows operators to work faster, covering more ground in less

time. It also reduces waste in processes like planting or irrigation by improving accuracy.

- **Accurate Steering Control:** Enhanced control over the tractor reduces the likelihood of accidents, improving safety in the field.
- **Increased Tire Durability:** Power steering contributes to better tire performance and longevity by improving contact with the ground and preventing issues such as tire skidding.

Difference Between Power Steering And Normal Steering

Power steering helps you steer with less effort and uses special assistance, while normal steering requires you to use your muscles to turn the wheel without any help. Learn more about power steering vs normal steering below:

Aspect	Power Steering	Normal Steering
Effort Required	Less effort is needed to turn the steering wheel.	More physical effort is required to steer.
Maneuverability	Provides smoother and more precise control.	Control may feel heavier and less precise.
Operator Fatigue	Reduces operator fatigue during long hours.	Operators may experience more fatigue.
Accessibility	Easier for operators of varying physical abilities.	This may pose challenges for some operators.
Cost	Generally higher cost due to added components.	Lower cost as it typically lacks power assistance.

Choosing The Right Power Steering System For Tractors

When selecting the right power steering system for tractors, factors such as brand, model, and price play an essential role. Leading options in the market include power steering systems from well-known brands like Mahindra, Swaraj, Sonalika, and John Deere. These brands offer a wide range of models, each with different power steering configurations and price points.

For example, prices for Swaraj 735, 834 XM, and Eicher 380 may vary depending on the specific model and its features. Similarly, Massey Ferguson models such as 1035 DI, 241, and 245 are available with power steering options, each priced according to its specifications. Other models like Mahindra 275 DI XP Plus and Eicher 242 also offer power steering, catering to different needs and budgets. Prices are influenced by factors like horsepower, technological features, and additional functionalities, so it's crucial to research and compare these options before making a decision.

The different types of steering gears are as follows:

1. Worm and sector steering gear.
2. Worm and roller steering gear.
3. Cam and double lever steering gear.
4. Worm and ball bearing nut steering gear.
5. Cam and roller steering gear.
6. Cam and peg steering gear.
7. Recirculating ball nut steering gear.
8. Rack and pinion steering gear.

Recirculating Ball Steering Gear

This type of steering gear is commonly used in Indian tractors. At the lower end of the steering column, a worm gear is present, with a box-shaped nut clamped onto it. Numerous ball bearings circulate between the worm gear and the nut. When the steering wheel is turned, the nut moves vertically. This movement is detected by the sector of the pitman connected to the nut, transferring the motion into the rotational movement of the pitman. Drop arms are mounted on the pitman shafts.

Worm and Roller Type Steering Box

In case of worm and roller steering, the worm at the lower end of the steering column is in the form of a cam. There is a roller which follows the shape of the worm. The roller is a part of the pitman. As the roller follows the cam when the steering column is turned, the motion is transferred to the pitman and to the drop arms. An exploded view of the worm and nut steering is given in

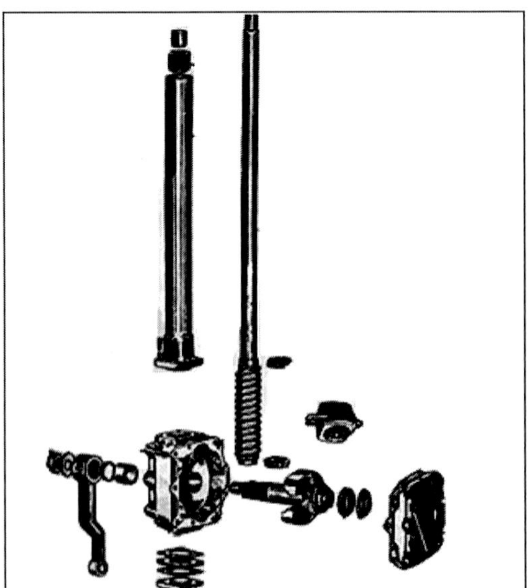

Figure: 13.4 Blow-up of a Recirculating Ball Steering Box

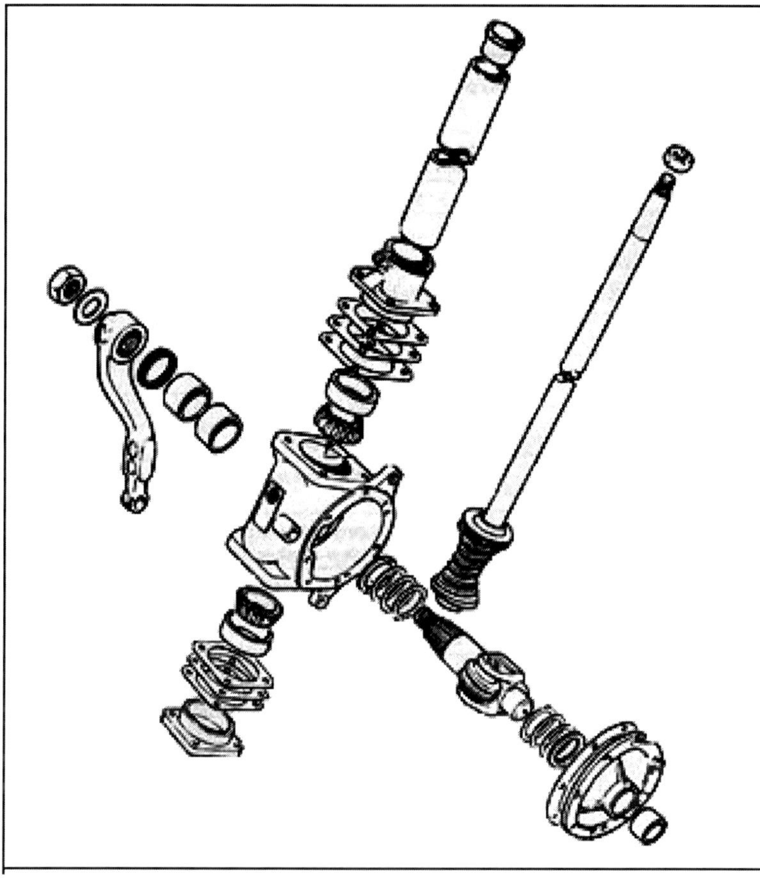

Figure: 13.5 Blow-up of a Wrom and Nut Steering

Worm and Sector Type Steering Box

In this type of steering box, the steering worm on the steering column rotates a steering gear sector that meshes with the worm. The gear sector then rotates the pitman on which it is mounted. The pitman is further connected to the steering linkage, which steers the wheels. This method shows how the rotation of the steering column is converted into the rotation of the pitman using the worm and sector steering system.

Rack and Pinion Type Steering Box

In a rack and pinion steering gear, a pinion is attached at the end of the steering shaft. When the steering wheel is turned, the pinion gear spins, moving the rack – left or right, depending on which way the steering is turned. The rack forms the part of the tie rod with steering spindle at its ends which push or pull the steering links for steering the wheels. Fig shows the arrangement for a rack and pinion steering box

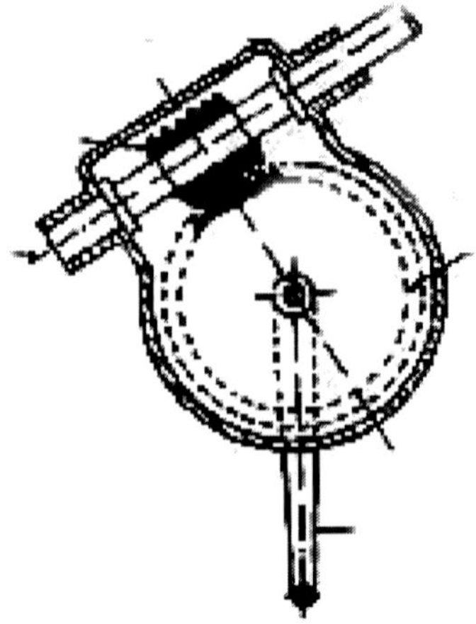

Figure: 13.6 Wrom and Sector Steering Box

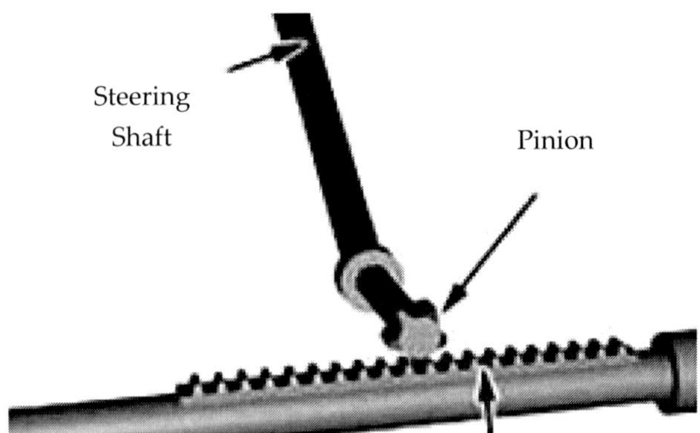

Figure: 13.7 Rack and Pinion Steering Box

The recirculating ball mechanism has the advantage of a much greater mechanical advantage, so that it was found on heavier vehicles while the rack and pinion was originally limited to smaller and lighter ones.

HOW DOES TRACTOR STEERING SYSTEM WORK?

The tractor steering system is responsible for controlling the direction of the vehicle by transferring the driver's input to the wheels, allowing for efficient navigation, especially in agricultural fields. The system works by converting the rotational motion of the steering wheel into linear motion, which then turns the tractor wheels.In most modern tractors, the steering system is typically a recirculating ball type or a worm and sector type, both of which are designed to offer precise and reliable control. The steering mechanism begins with the operator turning the steering wheel. The steering wheel is connected to the steering column, which transmits the rotational movement to the steering box. This steering box contains various components, such as the worm gear, sector gear, and pitman arm, that work together to steer the tractor.

In a recirculating ball steering system, as the operator turns the steering wheel, the movement is transferred to the worm gear at the lower end of the steering column. The worm gear meshes with a sector gear, and this interaction causes the sector gear to rotate. Attached to the sector gear is the pitman arm, which is mounted on a shaft. The movement of the sector gear moves the pitman arm in a linear direction, translating the rotational motion of the steering wheel into a lateral force that turns the wheels.Similarly, in a worm and sector type system, the steering worm at the bottom of the steering column rotates the steering gear sector. This gear meshes with the worm, and the sector gear, in turn, rotates the pitman arm. The pitman arm is then connected to the steering linkage, which further transfers this motion to the wheels. This design allows for precise control over the steering of the tractor and ensures the vehicle responds accurately to the operator's commands.

The power steering system is commonly used in modern tractors, especially those used for larger farming operations. Power steering reduces the effort required by the operator to turn the steering wheel, making it easier to maneuver the tractor. This system uses hydraulic pressure to assist the steering process. When the operator turns the steering wheel, hydraulic fluid is directed into a cylinder attached to the steering mechanism. The pressure from the hydraulic fluid provides additional force, aiding the operator in turning the wheel with minimal effort. This makes the tractor more responsive and easier to handle, even in challenging conditions such as working in mud or on uneven terrain.The tractor steering system is also equipped with a steering linkage, which includes various rods, arms, and joints. These components work together to transfer the motion from the pitman arm to the wheels. The steering linkage is designed to ensure that both wheels turn in unison, providing stable and predictable handling. In addition to the primary steering components, some tractors also have adjustable features that allow the operator to fine-tune the steering, further improving maneuverability.

In total,the tractor steering system is a complex yet vital component that ensures the tractor can be steered safely and efficiently. The integration of mechanical and hydraulic systems provides both precision and ease of use, allowing operators to work effectively across a wide range of terrains and agricultural tasks.

Technological Advancements in Tractor Steering

Modern tractors often feature GPS-guided steering systems that enhance precision during field operations. These systems use GPS technology to automatically steer the tractor, improving accuracy in tasks like planting, harvesting, and spraying.

14

Steering in Power Tiller

INTRODUCTION

The hydraulic power steering system operates by using pressurized power steering fluid to reduce steering effort. The system is powered by a pump, driven by an engine accessory drive or serpentine belt. The pump sends high-pressure fluid through a hose to the input side of the power steering gear-control valve. The high pressure applied to the control valve assists the driver in turning the front wheels. The power steering fluid is stored in a reservoir, and the fluid returns from the steering gear at a much lower pressure to maintain the proper level in the reservoir.

In a hydraulic power steering system, the fluid is continuously pumped, and the system is influenced by engine speed. High RPM results in high-pressure output, while lower RPM causes low-pressure output. To ensure constant pressure, the power steering pump uses a bypass valve, which regulates the pressure regardless of the engine speed. This helps maintain consistent performance.

ELECTRO-HYDRAULIC POWER STEERING

These systems employ a brushless electric motor to drive the hydraulic power steering pump instead of an engine driven accessory drive or serpentine belt. This system operates the same and provides the same feel of a conventional hydraulic power steering system.

Electric Power Steering or Motor-Driven Power Steering

This system eliminates hydraulics and pressurized fluid from the system. When the driver turns the steering wheel, a brushless bi-directional permanent magnet motor connected to the steering gear or steering column turns the front wheels. Sensors detect which way the steering wheel is turned to assist the movement of the steering gear in the correct direction. No matter what kind of system you have, power steering only acts on the steering gear when the driver turns the wheel.

Features

- High efficiency, long service life.
- Compact and convenient low pressure drops & steering torque ports available to DIN.
- ISO or SAE Size 50 to 400cc/rev available with built-in value functions shock.
- Inlet check suction and relief valves.
- According to European & US standards.extensively used in forklift.
- Tractor combines.
- And loaders.

DOG CLUTCH

A dog clutch is a type of clutch with square jaws designed to drive a shaft in either direction, commonly used in power tillers. Unlike friction clutches, which rely on friction to engage, dog clutches connect two rotating components through interference.

Figure: 14.1 Dog Clutch

The design of the clutch ensures that one part pushes the other, causing both to rotate at the same speed without slipping. This makes dog clutches ideal where slippage is undesirable, or where the clutch does not need to control torque. Because dog clutches do not rely on friction, they are less prone to wear compared to friction clutches.Dog clutches are often used in manual automotive transmissions to lock different gears to the rotating input and output shafts. A synchromesh system is typically employed to ensure smooth engagement by synchronizing the shaft speeds before the dog clutch is engaged.

A good example of a simple dog clutch can be found in a Sturmey-Archer bicycle hub gear, where a sliding cross-shaped clutch is used to lock the driver assembly to different parts of the planetary geartrain.

Advantages

1. No-slip or friction involved after the engagement of the clutch.
2. Heat generation is much less as no friction occurs.
3. Cost is less than friction clutch of similar torque capacity.

Disadvantages

1. Shock is caused while engagement of clutch at any speed.
2. Cannot be engaged at high speeds.
3. When both the diving and the driven shaft is at rest, some relative motion is required to engage.

FLUID COUPLING

A fluid coupling consists of a driving member and a driven member, typically an impeller with radial vanes, all enclosed within a casing. The coupler is mounted on the engine crankshaft and filled about three-quarters with an appropriate oil. A spring-loaded sealing ring ensures the driven shaft is oil-tight.

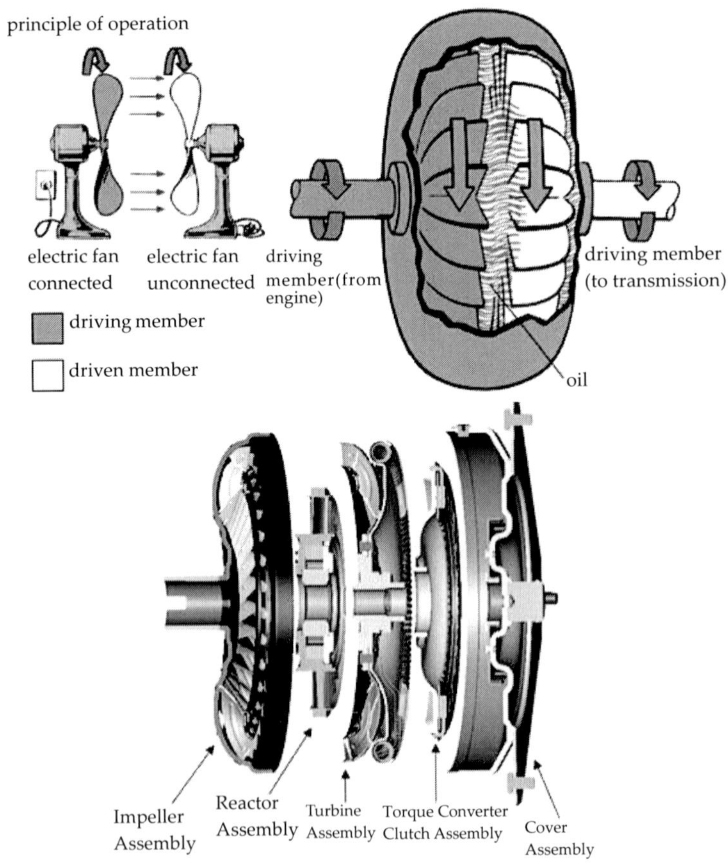

Figure: Fluid Couping

When the crankshaft rotates, centrifugal force pushes the oil from the center to the outer edge of the impeller, which increases the oil's velocity and energy. The oil then enters the runner at the outer edge and flows toward the center, causing the runner to rotate.

As long as the impeller and runner are rotating at different speeds, the oil continues to circulate uniformly. However, once the impeller and runner reach the same speed, oil circulation ceases. The fluid coupling does not increase the applied torque but ensures that the torque is transmitted evenly.

- The main features of fluid coupling are:
- Absorption of shock and vibration
- Smooth starting and
- Easy operation

15

Hydraulic System of Tractor

INTRODUCTION

Hydraulics is the study of liquids and their behavior, specifically focusing on the use of pressurized fluid to generate power in engineering systems. Hydraulic systems operate by applying pressure to a small amount of fluid, which in turn generates significant power. The development of hydraulic technology has enabled major advancements, such as lifting heavy loads and rotating large shafts with minimal effort through mechanical linkages.

Compared to mechanical and electrical systems, hydraulic power systems offer greater flexibility and produce more power relative to their size. They also provide quick and precise responses to control inputs. As a result, hydraulic systems are widely used in modern applications such as aircraft, automobiles, heavy industrial machinery, and various machine tools. This article will focus on the hydraulic systems used in tractors.

HYDROSTATIC TRANSMISSION

Hydrostatic transmissions rely entirely on hydraulics, using the power of oil to transmit energy. The system consists of two main components: a variable displacement pump and a hydraulic motor. The pump's displacement is controlled by a movable swash plate, which adjusts the piston stroke. Essentially, the system operates as a variable-displacement hydraulic pump driving a fixed-displacement hydraulic motor.

The key benefits of hydrostatic transmissions include the ability to infinitely vary ground speed and rapidly change directions. They are also known for their reliability, with self-protecting features that prevent operator abuse. Additionally, foot pedal-controlled systems have a built-in safety feature that allows for a controlled stop simply by lifting the foot from the pedal. However, hydrostatic transmissions have some drawbacks, including high cost, sensitivity to contamination, and slight power loss at the PTO shaft. Additionally, when parking on a slope, it's important to apply the parking brake to prevent unintended movement.

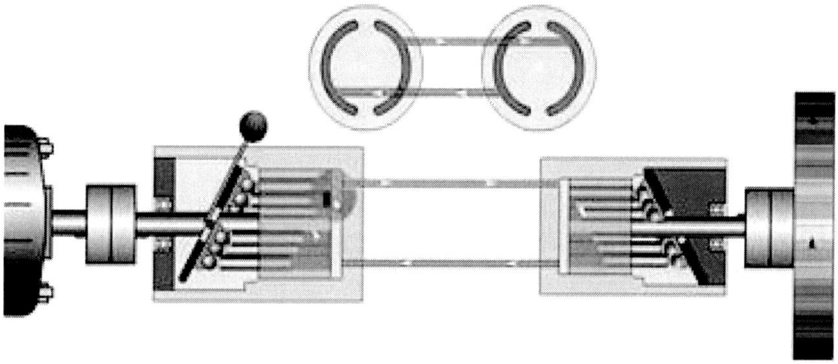

Figure: 15.1 Hydrostatic Transmission

Hydrostatic transmissions are ideal for turf mowing applications or tasks requiring frequent speed and direction changes within a limited area. They are commonly found in riding lawnmowers, lawn tractors, and other equipment needing continuously variable control.

A hydraulic automatic transmission consists of several key components:

- **Fluid Coupling Or Torque Converter:** This hydraulic device connects the engine and transmission, replacing a mechanical clutch. It allows the engine to continue running while at rest without stalling. Additionally, the torque converter provides variable torque multiplication at low engine speeds, improving acceleration when starting from a stop.
- **Planetary Gear Set:** A compound planetary set that utilizes hydraulic servos to actuate its bands and clutches, offering two or more gear ratios.
- **Valve Body:** Serving as the hydraulic control center, the valve body receives pressurized fluid from the main pump, which is driven by the fluid coupling or torque converter. This pressure is regulated and directed through a network of valves, check balls, and servo pistons. These valves use the hydraulic pressure to control which gear ratio is selected in the planetary gear set, adjusting automatically as the vehicle's speed and engine conditions change. In modern transmissions, these valves are often controlled by electro-mechanical servos managed by the Engine Management System or a separate transmission controller.

HYDRAULIC CONTROL SYSTEM

It is a mechanism in a tractor to raise , hold or lower the mounted implement or semi-mounted equipments by hydraulic means. All tractors are equipped with hydraulic control system for operating three point hitch of the tractor. Hydraulic system works on PASCAL's Law which states that pressure applied to an enclosed fluid is transmitted equally in all directions.

Basic Components of Hydraulic System

- Hydraulic pump
- Hydraulic cylinder and piston
- Hydraulic tank
- Control valve
- Safety valve
- Hose pipe and fittings
- Lifting arms

The hydraulic pump draws up oil from the oil reservoir and sends it to the control valve under high pressure. From the control valve, the oil goes to the hydraulic cylinder to operate the piston, which in turn, raises the arms. The implements attached with the arms are lifted up.

There is a special tank for hydraulic oil that is separate from the transmission chamber:

- **Hydraulic Pump:** They are of several types, such as gear pump, plunger pump, vane pump, and screw pump. Gear pumps are widely used in tractors because they can flow a large amount of oil in comparison to a plunger pump. The oil pressure in the pump varies from 150 to 200 kg/cm^2.
- **Hydraulic Cylinder:** It is a bigger cylinder fitted with a piston and a connecting rod, also called a ram cylinder. The piston moves in the hydraulic cylinder and causes reciprocating motion in the cylinder. It is the connecting rod that transmits power from the piston to the lifting arms. Although the lifting arm is raised by hydraulic pressure, it is lowered by its weight.
- **Hydraulic Tank:** It is used for storing hydraulic oil. In some tractors, the transmission chamber itself works as a hydraulic tank and the same oil is used for both transmission and hydraulic systems.
- **Control Valve:** It controls the movement of hydraulic oil to have desired direction, magnitude, and speed for lifting.
- **Oil Filter:** It is a small filter located at a convenient position in the passage of the oil.

Challenges that Hydraulic Parts have to Endure

Agriculture and its associated industries are undergoing constant evolution. With the latest advancements in science, agro-based industries are emerging in previously untapped areas. To meet land use standards, these regions often require significant topographical changes. Heavy machinery, including trucks and tractors, must operate in challenging climatic conditions, within tight timeframes, to bring these sites up to operational standards.

These machines rely on heavy-duty engines, which face intense workloads. To maintain engine performance and prolong lifespan, friction within the engine must be minimized. Hydraulic technology plays a crucial role in enhancing engine robustness, as it helps drive the crankshafts and ensures sustained high power delivery.

At Atrac, we are committed to anticipating the future needs of the agriculture and allied industries. Our advanced manufacturing and testing facilities enable us to produce high-quality hydraulic components, including camshafts, control arms, control valves, lift rams, and pistons. These components not only adhere to industry standards but also meet Original Equipment Manufacturer (OEM) specifications. Trusted in the field, these parts have proven their reliability and uphold our commitment to quality and assurance.

16

Tyres and Rims

INTRODUCTION

Tyres serve as the essential contact points between your vehicle and the ground. Without tyres, the power generated by the engine would be ineffective in producing movement. This power transfer is known as traction, and the effectiveness of a tyre's interaction with the surface significantly influences how much of the engine's power can be utilized. Traction becomes even more critical when operating heavy machinery, particularly for farm and field applications. Tractors, for instance, are built for rugged and demanding conditions, requiring more traction compared to regular tubeless car tyres.

A tyre consists of a torus, made of a complex, flexible, reinforced composite material that surrounds a steel hub and is filled with air under pressure. The tyre's rolling tread usually features expanded ribs. The design and shape of the torus, as well as the construction materials, are chosen based on the specific application of the tyre. Generally, a tyre is characterized by its construction type, mounting details, ply rating, and other key dimensions, as illustrated in relevant diagrams .

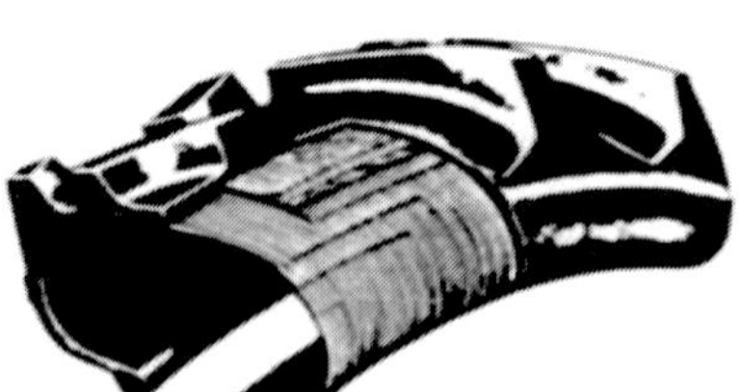

Radial Ply

Radial Ply

Figure: 16.1 The Construction Tyre

TIRE DEVELOPMENT

The steel-wheeled tractor was replaced by vehicles fitted with pneumatic tires in the 1920s and 1930s. Harvey Firestone (1868-1938) was instrumental in developing reinforced, "low-pressure" tires that were suitable for agricultural field work. World War II resulted in shortages of natural rubber previously used for tire production. This accelerated the research and development of synthetic rubbers such as styrene and latex. The synthetic rubbers produced had better wear-resistant properties than the original rubber tires. Synthetic rubbers that are commonly used today include polyurethane, neoprene, polybutandiene, and butyl. Natural rubber may still be used for some specialized applications.

Different Types of Tractor Tyres

Tractor tyres come in various types designed to suit different applications and terrains. Below are some of the most common types of tractor tyres:

1. **R-1 Tires:** These are standard agricultural tyres designed for general farming use, ideal for plowing and tillage work.
2. **R-1W Tires:** These tyres are tailored for wet conditions, featuring larger lugs to improve traction in muddy or slippery environments.
3. **R-2 Tires:** Designed for intermediate to heavy-duty tasks, R-2 tyres are typically used in row crop farming.
4. **R-3 Tires:** These tyres provide excellent traction in muddy conditions and are commonly used for heavy-duty tasks in orchards and vineyards.
5. **Flotation Tires:** These tyres are engineered for maximum flotation on soft or uneven terrain, making them suitable for fields with high soil moisture.
6. **Industrial Tires:** These tyres are built for use on construction machinery and provide strong traction on hard surfaces.
7. **Turf Tires:** Turf tyres are intended for lawn and garden tractors, designed to minimize damage to delicate lawns and landscaping.

These are some of the most common types of tractor tyres available. When selecting tyres, it's essential to consider the specific needs of your tractor and the tasks you'll be performing.

TIRE FUNCTION

The agricultural tyre must fulfill the following functions:

1. Support the vehicle and its load while maintaining low ground pressure.
2. Absorb shock loads and cushion the vehicle against minor surface irregularities.
3. Provide traction (and braking).
4. Ensure proper steering and directional stability.
5. Resist abrasion from various surfaces it operates on.

CATEGORIES OF TIRE APPLICATIONS

The application of tyres as traction elements can be divided into four broad categories, each with unique demands:

- **Construction:** In projects like dams, waterways, and highways, large quantities of earth and rock are moved. Such operations require tyres capable of handling high speeds (40 to 80 km/h), long hauls (up to 16 km), and heavy loads (up to 75 m3 capacity).

- **Logging, Mining, and Petroleum:** These industries require tyres with high flotation and load-carrying capacities, especially for heavy-duty equipment like mobile cranes and power plants.
- **Military Operations:** Vehicles in military transport must operate across various surfaces in cross-country environments, where reliability is critical.
- **General Transportation in Developing Areas:** Vehicles used in regions without sufficient infrastructure must have tyres that offer flotation and mobility under heavy loads without requiring extensive road or track maintenance.

TIRE SIZES

The growth of off-road operations has resulted in a diverse range of tyres to meet varying service requirements:

- **Number of Sizes:** Tyres have become larger in both cross-section and rim diameter, allowing for higher loads per tyre without compromising flotation.
- **Conventional vs Wide Base:** There are two types of tyres: conventional and wide base. Figure illustrates the differences. The width of the cross-section can be increased with a wider rim without changing the overall diameter or rim size. This reduces inflation pressures and improves traction, floatation, and resistance to damage from stones and other debris.
- **Low Section-Height Tires:** The wide base concept extends to low section-height tyres. Figure compares two tyre types. The low section height allows for a wider cross-section, improving flotation without increasing diameter or weight.
- **Single vs Dual:** A variation of the low-section height principle is using larger single tyres instead of dual tyres. Figure shows this comparison. Although the ground contact area is reduced, experience suggests that floatation and mobility are improved without compromising load capacity.
- **Tread Pattern:** Off-highway operations do not always require the same level of traction. Therefore, different tread designs are used for varying degrees of tractive effort.

TREAD

The key factors related to tires include:

(i) Size (diameter and width), which affects their tractive capacity and rolling resistance. (ii) Strength, indicated by the ply rating, which determines the tire's pressure tolerance and weight-carrying capacity. This, in turn, influences both the tractive capacity and rolling resistance. (iii) Tread pattern, which, combined with the surface characteristics, impacts how the tire engages or contacts the surface.

Power losses at the wheel/surface interface can be significant, particularly on soft surfaces, leading to lower efficiency. As a result, the power available at the tractor's drawbar may be considerably less than the engine's output. Therefore, selecting the appropriate tires and managing the weight placed on them are essential for optimizing the tractor's overall performance.

Figure: 16.2 Ground Drive Elements

Various types of wheels and / or tyres may be used on the tractor, depending mainly on the surface on which it is working. For the following conditions, the tyres or wheels indicated are recommended as shown in Figure .

	Surface	Tread form
(a)	Hard surfaces such as roads	Large area, shallow tread with 'high' pressure
(b)	Normal agricultural work, dry soil	Heavy, intermediate depth tread
(c)	Soft, wet agricultural soils	Deep tread
(d)	Lawns, low sinkage is required	Wide, low pressure
(e)	Dry soil, heavy loads as in earth moving	Tracks, as on a "crawler" tractor
(f)	Saturated puddle soils	Metal cage, with angled lugs, alone or as extensions to normal tires

Maintaining Ideal Air Pressure

Tractor tyre life is generally longer than an average car or bike tyre. However, with certain precautions and checks, you can get more mileage out of your tractor tyres and hence save on long-term costs of maintaining the vehicle. As mentioned above, the first step is to pick the right tractor tyre. Then comes keeping a regular check on air pressure of the tractor tyres – front and rear. Maintaining ideal air pressure helps in avoiding uneven tread wear and increases the fuel efficiency of your tractor.

How Long do Tractor Tires Last?

The lifespan of a tractor tire depends on various factors such as the type of tire, the type of work it is used for, the loads it carries, and the terrain it operates on. Here are some general estimates of the lifespan of tractor tires:

- **Bias-Ply Tires:** Bias-ply tires are typically more durable than radial tires, and can last up to 10,000 hours or more, depending on use.
- **Radial Tires:** Radial tires tend to have a longer lifespan than bias-ply tires, and can last up to 15,000 hours or more, depending on use.
- **Dual Tires:** Dual tires, which are a pair of tires mounted on the same wheel, typically have a longer lifespan than single tires, as the weight of the load is distributed over two tires.

RIM

The rim plays a crucial role in supporting the weight of the tractor and transferring engine torque to the tyre, which then transmits the force to the ground. It needs to be appropriately sized, considering the tractor's weight, its implements, and the engine power. Failure to properly size the rim could lead to a risk of it breaking under pressure. The wheel consists of two main components: the rim, which is a ring that supports the tyre, and the disc, which allows the wheel to be fixed to the hub. Both parts may vary in shape and dimensions. The disc can either be removable or welded to the rim ring. While welding does not affect the rim's efficiency, it limits the flexibility in configuration compared to a removable disc.

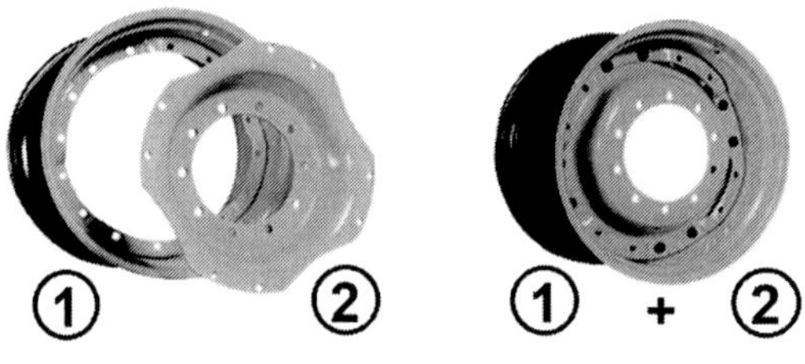

Figure: 16.3 The Rim Ring is Designed to Accommodate the Tyre,The Disc Allows the Rim to be Fixed to the Hub

How to Establish the Exact Dimensions of an Agricultural Rim?

To find out the exact dimensions of your rims, you must measure the two main dimensions:

Measure the height or diameter , taking care to measure from the tyre bead seat and above all not from the outer lip of the rim. This rim diameter is the reference point generally accepted by everyone, to designate a rim of 20 inches, 32 inches, 42 inches, etc.

The second dimension is the width of the rim which is measure from the inside of the outer lip, at the level of the bead seat. This second measurement is also expressed in inches.

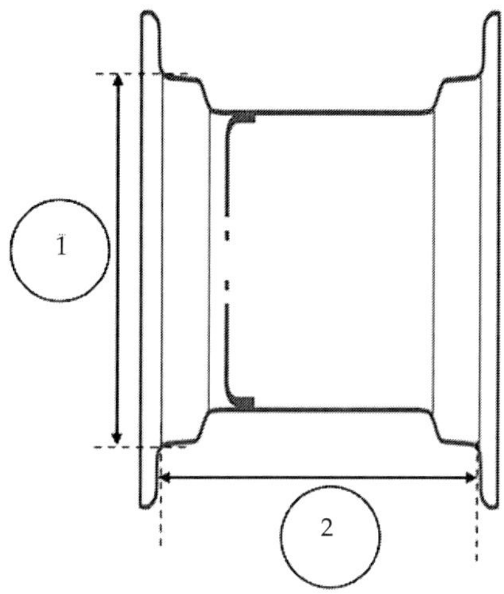

Figure: 16.4 Height of the Rim Diameter in Inches,Width of the Rim in Inches

Explanations on the Role of the Rim

The rim plays a crucial role as it supports the weight of the tractor and transmits engine torque to the tire, and subsequently to the ground. It must be properly sized to match the tractor's weight, including its implements, and the engine's power. If the rim is not appropriately sized, it risks breaking under pressure. The engine transmits torque to the rim, subjecting the metal to a turning force that is absorbed by the rim's thickness. If the metal is too thin or not strong enough to withstand the combined forces of the engine torque and the tractor's weight, it may crack.

Figure: 16.5 Cracked Rim Disc

In which Case does the Tyre Rotate on the Rim?

What is important to keep in mind is that the rim disc is bolted to the axle at the level of the hub, but this does not apply to the wheel, which is fixed to the rim merely by the pressure applied to the contact area between the tyre bead and bead seat on the rim.

The purpose of the knurling under the bead is to prevent the tyre slipping on the rim. However, if this does occur nonetheless, the knurling will begin to destroy the bead seat rubber, then end up damaging the bead wire, leading to obligatory replacement of the tyre.

Tyre slip may occur principally if there is too much torque or if the tyre is not correctly mounted (incompatible tyre and rim dimensions).

Here are the reasons that cause tyre slip:

- Tyre inside diameter variance, or tyre size not suitable for the rim.
- Rim outside diameter variance, or rim seat not up to tyre specifications.
- Rim polished or painted, often to prevent rust. In reality, rust is not a major problem on the knurling if it is solid and clean, it can even help the bead to catch on, which is not the case if the knurling is saturated with paint.
- Use of improper lubricant during mounting of the tyre (vegetable or animal oils may be used but never antifreeze, petrol, ether, silicon, oil, grease or any other petroleum-based lubricant, nor anything that stays slippery after use).
- Knurling that has become worn or rounded over time.
- Insufficient inflation pressure which reduces the amount of contact area between tyre and rim.
- Rim slip may also increase as a result of high torque demand.

What are the Differences Between Rim Models?

There are many types of rims, but there are 3 profiles which are most commonly used: W, DW and TW rims, these letters corresponding to the specific shape of the rim on the ring just before the bead seat position. This designation is totally independent of the type and shape of the central disc which may be bolted, partially welded or fully welded, with different hub diameters.

The W, DW, TW...etc. rim design parameters were established by standardisationorganisations (such as ETRTO), while no provisions of this type have been taken for discs, whose characteristics vary considerably from one tractor or wheel manufacturer to another.

The W rim

- **The Single Well Shape Rim, W:** This is an old model which is starting to be replaced by the double well shape rim for large or medium-sized tyres, because for the W model, the mounting of the second bead wire is more difficult. This model is less resistant than the more recent designs. However the W rim is still suitable for small sized rims, for which there are no DW or TW models on the market.

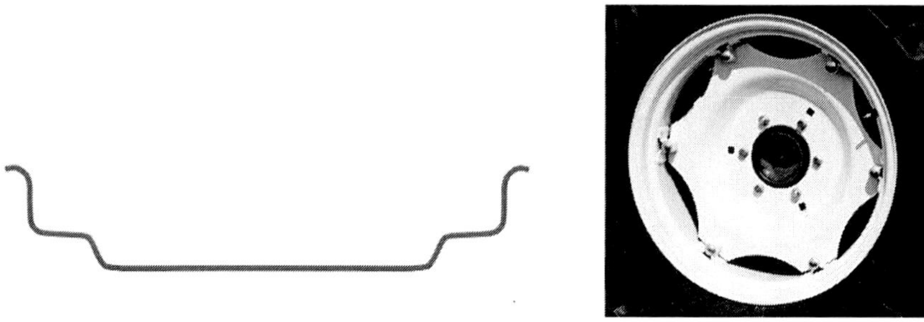

Figure: 16.6 Jante Creuse Simple W (Anglais W = Single Well Shape)

The DW rim

- **The Double Well Shape Rim, DW:** this design is the most commonly used at present, because it is more robust than the W model and is easier to use during mounting and setting up operations. It allows more possibilities for the disc.

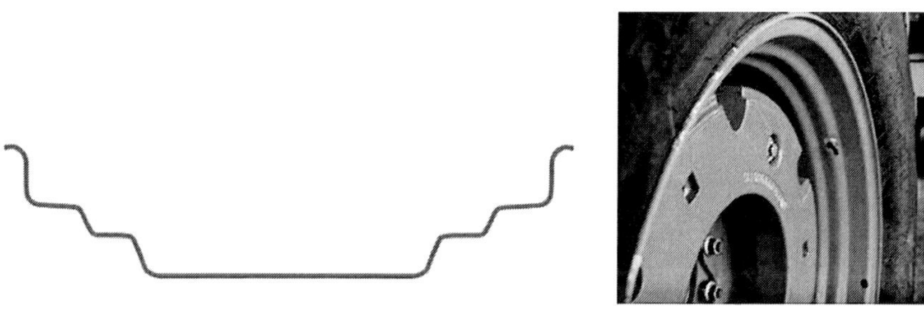

Figure: 16.7 Double Well Shape Rim (DW)

The TW rim

The TW shape is a design that is growing in popularity; it offers extra structural resistance and its shape makes mounting easier as the tyre slides more progressively over the three steps when putting the bead in place.

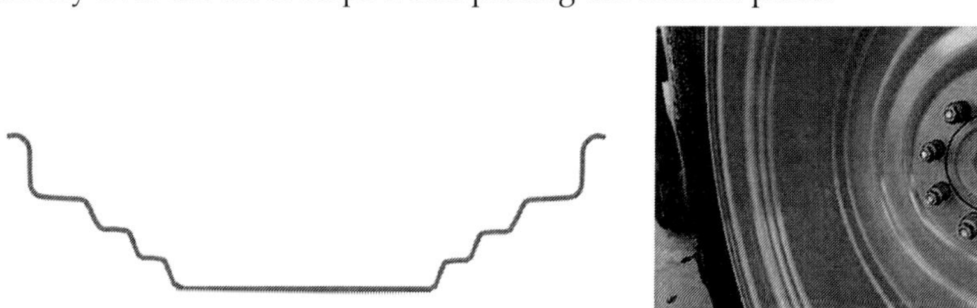

Figure: 16.8 Triple Well Shape Rim (TW)

Why change or swap rims?

Changing rims can enhance productivity by enabling the use of larger tires with greater air volume, which allows for heavier implements without causing soil compaction. The following options are available when considering rim changes:

- **To Change Tire Dimensions:** Opting for taller or wider tires lets you work with heavier loads and achieve better yields, while also protecting the soil from compaction.
- **Reduce the Rim Diameter Without Altering Tire Bulk:** This modification increases the air volume without expanding the tire's width, enabling it to bear more weight.
- **Adjust Tractor Track Width:** Depending on the crop or task, such as inter-row cultivation, rims with adjustable discs can modify the track width. Variable offset rims also allow the user to fine-tune the track, and welded rims can offer a positive or negative offset by reversing them.
- **To Adapt a Second Rim for Dual Tires:** Adding another rim allows for dual tire setups, improving traction and load distribution.
- **To Switch from Diagonal (Bias Ply) to Radial Tires:** This change can improve tire performance and longevity, offering better grip and efficiency.

FUNCTIONS OF TRACTOR WHEELS

Generally, all tractor wheels perform the following functions:

- They provide steering and directional stability.
- The wheels cushion the vehicle from the abrasive action of operating surfaces.
- They support the vehicle and associated loads
- The treads on the tires provide traction
- They absorb road shocks and protect the tractor from minor surface irregularities.

17

Daily, Weekly and Monthly Maintenance Schedule

INTRODUCTION

Maintenance scheduling involves planning, organizing, and coordinating the tasks required to ensure equipment operates efficiently. A maintenance schedule is a comprehensive plan that specifies the timing and frequency of maintenance activities for machinery, equipment, and facilities within an organization. Its primary goal is to maintain optimal functionality, minimize downtime, and extend the lifespan of assets. The schedule typically includes an inventory of assets, details of maintenance tasks, their frequency, assigned personnel, required documentation, and identification of resources necessary to complete the maintenance work.

IMPORTANCE OF MAINTENANCE SCHEDULING

A well-structured maintenance schedule is essential for ensuring efficient and reliable operations within the manufacturing industry. By using an effective maintenance scheduling system like MaintWiz CMMS, you can enjoy the following advantages:

- **Improved Equipment Performance:** Performing routine maintenance enhances the reliability of your equipment, which in turn leads to better overall performance and increased productivity.
- **Enhanced Safety:** Scheduled maintenance includes regular safety inspections, helping to reduce the likelihood of accidents and contributing to a safer working environment.
- **Minimized Downtime:** Regularly scheduled maintenance reduces the likelihood of unexpected breakdowns and helps to cut down on unplanned downtime, ensuring the smooth functioning of equipment.
- **Optimized Resource Utilization:** Scheduling maintenance tasks properly allows for the best allocation of resources, ensuring that tasks are completed based on priority and availability, which boosts operational efficiency.

- **Cost Savings:** By addressing potential issues proactively through a well-organized maintenance schedule, you can avoid costly emergency repairs and help extend the operational life of your equipment.

ASSESSING EQUIPMENT MAINTENANCE NEEDS

Before creating an effective maintenance schedule, it is crucial to assess the specific maintenance needs of your equipment. This process includes:

- **Determining Maintenance Frequency Based on Asset Priority:** Establish maintenance intervals based on the importance of the asset, factoring in manufacturer recommendations and the equipment's role in overall operations.
- **Conducting a Comprehensive Equipment Inventory:** Take stock of all assets that require maintenance and categorize them according to their criticality and priority to prioritize maintenance efforts.
- **Identifying Critical and Non-Critical Assets:** Focus maintenance efforts on critical assets that have a major impact on production, minimizing the risk of downtime for these key components.

TYPES OF MAINTENANCE SCHEDULES

There are various types of maintenance schedules, each tailored to meet the specific needs of an organization, its equipment, and operational goals. Here are some of the most common types of maintenance schedules:

Preventive Maintenance Schedule

Preventive maintenance involves performing scheduled activities regularly to prevent potential equipment failures. These activities include routine inspections, lubrication, filter changes, and part replacements that occur at defined time intervals, such as monthly or quarterly.

Corrective Maintenance Schedule

Corrective maintenance is carried out after a failure occurs, aiming to restore the equipment to its normal operating condition. This type of maintenance involves repairing or replacing components that are broken or malfunctioning, often after an unexpected breakdown or issue.

Predictive Maintenance Schedule

Predictive maintenance focuses on monitoring the condition of equipment to predict when maintenance should be performed. This is done by using monitoring tools like vibration analysis, thermography, and oil analysis, which help determine the optimal timing for maintenance activities based on the equipment's performance and condition.

Condition-Based Maintenance Schedule

Similar to predictive maintenance, condition-based maintenance relies on real-time data and ongoing monitoring of equipment to decide when maintenance is necessary. It involves tracking variables such as temperature, pressure, and wear levels and triggers maintenance actions only when specific thresholds are met.

Time-Based Maintenance Schedule

Time-based maintenance involves performing maintenance tasks at regular, predetermined intervals, regardless of the equipment's condition. This strategy ensures that machinery and systems are serviced at set times to prevent potential failures and maintain optimal performance.

Types of Maintenance Tasks

A maintenance schedule should encompass various types of maintenance tasks to ensure comprehensive asset management. These include:

1. **Preventive Maintenance Tasks:** Scheduled inspections, cleaning, lubrication, and other proactive measures to prevent failures and extend equipment life.
2. **Corrective Maintenance Tasks:** Reactive repairs and fixes carried out in response to identified issues or breakdowns.
3. Predictive Maintenance Tasks: Utilizing data and analytics to anticipate and address potential equipment failures before they occur.
4. Event-Based Maintenance Tasks: Specific maintenance actions triggered by events such as equipment reaching a certain usage threshold or a system malfunction.
5. Planned Vs. Unplanned Maintenance Activities: Balancing scheduled maintenance tasks with unscheduled, emergency repairs to maintain optimal equipment performance.

MAINTENANCE INTERVALS

The frequency at which maintenance tasks are performed plays a vital role in maximizing equipment performance and minimizing disruptions. Maintenance intervals can be determined based on:

- **Time-Based Maintenance Intervals:** Setting specific time intervals, such as daily, weekly, monthly, or annually, for conducting maintenance tasks.
- **Usage-Based Maintenance Intervals:** Aligning maintenance schedules with equipment usage metrics, such as operating hours, cycles, or output quantities.
- **Condition-Based Maintenance Intervals:** Monitoring the condition of equipment through sensors, data analysis, and diagnostics to schedule maintenance based on actual asset health.
- **Factors Influencing Maintenance Interval Determination:** Considerations such as equipment criticality, historical performance data, manufacturer recommendations, and industry best practices.

SCHEDULING TECHNIQUES

When developing a maintenance schedule, it is crucial to adopt effective strategies and tools to streamline the process and ensure optimal allocation of resources. Key factors to consider include:

- **Time-Based Scheduling:** This involves assigning maintenance tasks at predefined intervals, ensuring consistent inspections and upkeep to prevent unexpected issues.
- **Calendar-based scheduling:** Aligning maintenance activities with specific dates or events such as seasonal changes, holidays, or planned shutdowns ensures that all tasks are timely and organized.
- **Event-Based Scheduling:** Maintenance tasks are triggered based on certain events or conditions, such as equipment alerts, sensor readings, or system-generated triggers, ensuring proactive responses to issues.
- **Hybrid Scheduling for Optimization:** Combining multiple scheduling techniques can help maximize efficiency and ensure the best use of available resources by integrating time-based, event-based, and other methods.
- **Resource Availability Considerations:** It's important to factor in the availability of maintenance teams, spare parts, and equipment access to avoid scheduling conflicts or delays that could impact operations.
- **Utilizing Maintenance Management Software or Cmms:** Advanced tools like MaintWiz CMMS can automate the scheduling process, track maintenance activities, and generate reports, which help in analyzing performance and improving the maintenance process over time.

CREATING A MAINTENANCE SCHEDULE TEMPLATE

To ensure consistency and ease of use, it's beneficial to create a maintenance schedule template tailored to your organization's needs. The template should include:

- **Designing a Structured Schedule Format:** Organize the schedule with clear sections for asset details, task descriptions, frequencies, responsible parties, and any specific instructions.
- **Including Necessary Information:** Ensure relevant asset information, such as serial numbers, location, and historical data, is easily accessible in the schedule.
- **Using Digital Tools for Easy Schedule Management and Updates:** Employing maintenance management software like MaintWiz CMMS allows for efficient creation, editing, and sharing of schedules, ensuring real-time visibility and collaboration.

BALANCING PREVENTIVE AND CORRECTIVE MAINTENANCE

When establishing a maintenance schedule, it's crucial to strike a balance between preventive and corrective maintenance strategies. Consider the following:

- **Determining the Right Balance Between Preventive and Corrective Maintenance:** Assess the cost and impact of breakdowns versus proactive maintenance to find an optimal approach.
- **Considering Equipment Criticality and Reliability:** Focus preventive maintenance efforts on critical assets or those prone to failures, while corrective maintenance can address less critical equipment.

- **Incorporating Feedback from Maintenance Teams and Operators:** Encourage open communication and input from those directly involved in equipment maintenance to fine-tune the balance and optimize the schedule.

OPTIMIZING MAINTENANCE ROUTES AND WORK ORDERS

Efficiently organizing maintenance tasks can greatly enhance productivity and reduce travel time. Consider the following optimization strategies:

- **Grouping Maintenance Tasks by Location or Equipment Proximity:** Cluster tasks in a way that minimizes travel time and maximizes efficiency.
- **Minimizing Travel Time and Maximizing Efficiency:** Plan routes and schedules to minimize travel distances and avoid unnecessary backtracking.
- **Streamlining Work Order Generation and Assignment:** Utilize digital tools and CMMS software to automate work order generation, assignment, and tracking, eliminating manual paperwork and reducing administrative overhead.

CONTINUOUS SCHEDULE MONITORING AND IMPROVEMENT

Creating and maintaining an effective maintenance schedule is an ongoing process that involves regular monitoring and continuous improvement. The following steps are crucial for ensuring its success:

- **Regularly Reviewing and Adjusting the Maintenance Schedule:** Consistently evaluate the performance of the schedule and make necessary updates based on feedback, equipment performance metrics, and evolving operational needs to maintain its relevance and effectiveness.
- **Analyzing Maintenance Data and Performance Metrics:** Utilize data collected through CMMS software or other tracking tools to identify patterns, recurring issues, and inefficiencies. This analysis helps optimize maintenance intervals, refine tasks, and improve overall performance.
- **Incorporating Feedback from Maintenance Teams and Operators:** Actively engage with maintenance personnel and equipment operators to gather insights from their experiences. Their feedback can pinpoint inefficiencies, suggest practical adjustments, and contribute to enhancing the overall efficiency of the maintenance process

Benefits of a Well-Executed Maintenance Schedule

Implementing and maintaining a well-executed maintenance schedule with MaintWiz CMMS can lead to numerous benefits for your manufacturing operations:

- **Improved Equipment Reliability nd Uptime:** Regular maintenance minimizes unexpected breakdowns, improving equipment reliability and reducing costly production disruptions.
- **Enhanced Operational Efficiency:** Optimal scheduling ensures maintenance activities are conducted without interfering with critical production processes, maximizing overall operational efficiency.

- **Reduced Emergency Breakdowns And Associated Costs:** Proactive maintenance helps prevent emergency breakdowns, reducing the need for costly repairs, downtime, and potential safety risks.
- **Extended Equipment Lifespan And Reduced Capital Expenditure:** Regular maintenance and timely repairs can extend the life of your assets, reducing the need for premature equipment replacements and associated capital expenditures.

HOW TO MAKE A MAINTENANCE SCHEDULE

Creating an effective maintenance schedule involves systematically organizing activities to ensure all equipment is regularly reviewed, maintained, and replaced as needed. Follow these four essential steps:

List All the Equipment to Maintain

Begin by inventorying all equipment requiring maintenance. Conduct a thorough inspection to compile a complete and comprehensive list, ensuring no machinery or tool is overlooked. This step guarantees that every item needing maintenance is accounted for, forming a solid foundation for effective scheduling.

Identify Maintenance Activities

After listing the equipment, determine the specific maintenance activities required for each item. Research the needs of each piece of machinery, including routine inspections, lubrication, part replacements, or repairs. Prioritize equipment based on its usage frequency, return on investment, or repair costs, and create a checklist of maintenance tasks. This step simplifies the scheduling process and ensures all critical activities are addressed.

Determine the Frequency of Maintenance

Schedule inspections and maintenance based on the frequency required for each piece of equipment. There are likely activities that must be performed daily, while others can wait a week or month or can be done whenever required through a work order. These can include inspections, repairs and other maintenance tasks. Be sure to define the maintenance frequency for each item on the equipment list.

Define Maintenance Dates

Finally, to execute the maintenance schedule, the above information must be scheduled. This includes calendaring the maintenance schedule to ensure that every piece of equipment gets the maintenance it requires with the frequency to ensure it continues to work as expected. This maintenance schedule should be tracked and adjusted as needed.

PRINCIPLES OF MAINTENANCE PLANNING

Protect the Planner

Planners are separated from maintenance crews and placed into dedicated groups to concentrate on specialized planning techniques and future work. This separation protects the planning function by ensuring planners report to a different supervisor than the crew they plan for. Planners must not be used as field technicians to assist with ongoing work; instead, they should focus entirely on planning. While it may be tempting to involve planners in fieldwork during challenging situations, keeping them dedicated to their planning role ensures long-term efficiency.

Focus on Future Work

The planning team should solely focus on tasks that are yet to begin, ensuring a backlog of at least one week of pre-planned work for the maintenance team. This backlog supports the creation of a structured weekly schedule. Planners should not address immediate issues arising during active jobs; instead, supervisors or technicians themselves should resolve such problems. After completing a task, supervisors or lead technicians should provide feedback to the planning team, sharing insights about any encountered issues and suggesting improvements for future plans. This separation allows planners to stay focused and avoid distractions from other tasks, such as handling immediate equipment issues or assisting technicians, which can sidetrack their responsibilities.

Component-Level Files

Planners should maintain a simple, secure file system organized by equipment tag numbers at the component level, rather than the system level. This allows planners to gather and utilize equipment data from previous jobs to improve future work plans, particularly for repetitive tasks. Creating "mini-files" for each component after its first maintenance activity helps accumulate valuable data over time. These files should correspond to the component tag numbers used in the field, enabling planners to refine processes and enhance efficiency for recurring tasks.

Use Planner Judgment for Time Estimates

Planners should leverage their expertise and available data to develop realistic time estimates for work orders. Time estimates should reflect the actual duration a competent technician might require to complete a job under normal conditions. This principle emphasizes the importance of selecting planners from the organization's most experienced technicians. For instance, a planner with years of field experience may identify discrepancies in previous time estimates and provide more accurate predictions based on practical knowledge. This ensures that the time allocated for tasks aligns with realistic expectations.

Recognize the Skill of the Technicians

When creating job plans, planners must consider the skill level of the technicians who will execute the tasks. They should outline the scope of work and provide a general strategy, including preliminary procedures if none exist, tailored to the technicians' expertise. The amount of detail in the job plan should balance between providing sufficient guidance for less-skilled technicians and avoiding redundancy for highly skilled ones. For example, instead of drafting overly detailed plans, planners should focus on covering all work hours effectively. Utilizing existing procedures or notes in job plans ensures technicians can efficiently complete tasks while collaborating with planners on repetitive jobs to refine processes.

Measure Performance with Work Sampling

The effectiveness of planning and scheduling is measured primarily through "wrench time," which refers to the time technicians actively spend working on tasks without delays caused by waiting for assignments, tools, parts, clearances, or travel. Effective planning reduces on-site delays, while proper scheduling minimizes gaps between tasks. By focusing on reducing unnecessary downtime and optimizing wrench time, planners and schedulers contribute significantly to workforce efficiency and the successful execution of maintenance tasks.

Maintenance Planning Principles	Maintenance Scheduling Principles
1. Protect the planner	1. Job plans are needed for scheduling
2. Focus on future work	2. Scheduling and job priorities are important
3. Component-level files	3. Schedule based on the highest skills available
4. Use planner judgement for time estimates	4. Schedule for every available work hour
5. Recognize the skills of the techs	5. Daily work is handled by the crew leader
6. Measure performance with work sampling	6. Measure performance by schedule compliance

PRINCIPLES OF MAINTENANCE SCHEDULING

Job Plans are Essential for Scheduling

Effective scheduling begins with comprehensive job plans, which detail key elements like the number of technicians required, their minimum skill levels, estimated work hours for each skill, and job duration. These job plans provide the foundation for scheduling by answering critical questions, such as whether welding is needed, how many welders or assistants are required, and the specific roles involved. By addressing these details during the planning stage, schedulers can allocate resources efficiently and streamline the scheduling process.

Importance of Schedules and Job Priorities

A well-organized weekly schedule, aligned with job priorities, is crucial for boosting productivity. Weekly schedules allow crew supervisors to concentrate on current tasks without being distracted by the backlog. These schedules help coordinate maintenance and operations in advance, ensuring that tasks are properly prioritized. Assigning workloads based on these priorities ensures that critical tasks are completed first and sufficient work is distributed to maintain productivity.

Schedule Based on the Highest Available Skills

A one-week schedule should be developed for each crew, factoring in available technician hours, the highest skill levels, job priorities, and job plan details. By selecting tasks from the plant backlog based on priority and matching them with the crew's projected capabilities, schedulers optimize efficiency. Over time, technicians become accustomed to their weekly workloads, improving their overall productivity. This approach ensures that the most skilled individuals are matched to the appropriate tasks for maximum output.

Schedule All Available Work Hours

To maximize workforce utilization, the scheduler should assign tasks to cover 100% of the available work hours for the week. For instance, if a crew has 800 labor hours available, 800 hours of work should be scheduled. This approach prevents over- or under-scheduling and ensures that all available time is productively utilized. By fully aligning the workload with forecasted hours, the scheduling process minimizes idle time and optimizes crew performance.

Daily Work Managed by Crew Leaders

The crew leader or supervisor is responsible for creating and adjusting daily schedules based on the weekly plan, current progress, and any emergent high-priority tasks. Daily scheduling involves assigning tasks based on technician skill levels and work order requirements, as well as handling emergencies or rescheduling as needed. Since daily schedules are subject to the fluid nature of ongoing work, they require flexibility to accommodate task progress, inaccuracies in time estimates, and reactive maintenance demands.

Measuring Performance Through Schedule Compliance

The effectiveness of scheduling is measured by adherence to the weekly schedule and the resulting productivity. "Wrench time," or the time technicians spend actively working, is the ultimate indicator of workforce efficiency and scheduling success. By reducing delays within jobs through planning and between jobs through scheduling, the process ensures smoother task execution and enhanced workforce utilization.

How to Implement Maintenance Planning and Scheduling

Now that you understand the guiding principles of maintenance planning and scheduling, let's take a look at how to implement them. Start-to-finish implementation can be viewed in six phases:

- **Phase 1 - Setup:** This phase encompasses all the steps needed to ensure your organization is onboard with implementing maintenance planning and scheduling. You should have made your case to leadership by exposing the issue of low productivity, explaining how planning and scheduling can help solve that issue, calculating the value of productivity improvement, and presenting the results in the form of return on investment (ROI).
- **Phase 2 - Define and Analyze the Situation:** Phase two involves your team looking at your current situation and identifying problems currently faced in

maintenance execution. During this phase, you should have representation from all levels of the maintenance process — technicians, key managers or supervisors, and even representatives from procurement, finance and the warehouse. This workshop-like environment should outline the current maintenance planning and scheduling process.

- **Phase 3 - Develop and prepare for delivery:** Phase three involves planners and supervisors working to establish supporting documentation and process maps as well as defining in detail new processes, roles and responsibilities. You should also make any necessary changes to your computerized maintenance management system (CMMS) and develop training and coaching programs. Conducting a single overview training session followed by a role-specific training program is the most efficient way to go about training. This will prevent people from having to attend training sessions that don't pertain to them.
- **Phase 4 - Implement:** Once everything is in place, it's time to roll out the new maintenance planning and scheduling processes. The goal here is to embed the new standards and procedures into the daily routines of all those involved until they become the new normal. It's generally accepted to allow for a three-month coaching period, where individuals are assessed and receive help to close any gaps in performance. If you operate a shift system, six months should be sufficient. Remember, planners should only work on the processes, not in the processes.
- **Phase 5 - Review:** This is sometimes called the "close-out" phase. Here, you want to ensure the new maintenance planning and scheduling process won't disintegrate when the training and one-on-one time is over.
- **Phase 6 - Sustain:** This phase is considered "evergreen," as processes and procedures should always be improving. Be sure you have:
    - All performance metrics in place and review them in meetings, verifying that they are meeting long-term trends.
    - Clearly defined procedures or job plans for each technician performing certain tasks.
    - Ensure new technicians are properly trained on these job plans.
    - Standardized, up-to-date and easily accessibly documentation in place.

A set time for conducting process reviews to assess what is working and what isn't. This is also the time to go over how processes can be improved.

MAINTENANCE PLANNING AND SCHEDULING TOOLS

Many people consider planning to be a maintenance tool. While not solving all maintenance issues by itself, planning integrates with other elements and helps synchronize all aspects of maintenance. Let's take a look at some other tools that work together with maintenance planning and scheduling.

- **Work Order Systems:** A work order system is one of the most powerful tools a maintenance team can use. It acts as centralized, automated way to request and record work done within an organization. Work order systems are important because of the amount of work maintenance supervisors need to track. If a supervisor is responsible for eight technicians, each of those technicians might complete two or three tasks a day, totaling around 40 to 60 assignments in a five-day workweek. Work order systems give crews and supervisors a single method of communication. In addition to an integrated communication platform, a work order system lets supervisors organize workloads, assign tasks and track completion.

- **Equipment History and Data:** The equipment's history and data should also be considered a tool because this information helps you figure out the proper maintenance required based on up-to-date or even real-time data, as opposed to relying on memory or experimentation. Recording maintenance data over time essentially gives you a photo album of the life of the machine.
- **Maintenance Metrics:** Maintenance metrics refer to the measurements and scores of maintenance activities or results. They include the selecting, collecting, analyzing and presenting of maintenance data. The amount of work orders in your backlog is a common example of a maintenance metric. Metrics can help with everything from clarifying situations to planning for key performance indicators (KPIs), coverage, work type, schedules, compliance and backlog work hours.

TIPS FOR EFFECTIVE MAINTENANCE PLANNING AND SCHEDULING

Select an Experienced Maintenance Planner

A successful maintenance planner is often a senior technician who has extensive knowledge of the plant, a strong rapport with coworkers, and a deep understanding of maintenance planning principles. This individual should be skilled, resourceful, and familiar with the intricacies of the facility to make informed decisions.

Train the Planner Thoroughly

Proper training is essential for planners to effectively use work order software, extract data, and generate reports. This ensures they have access to the maintenance history of equipment and can create well-informed maintenance plans.

Separate Planning and Scheduling

Planning and scheduling are distinct functions and should remain separate. Planning involves identifying tasks, determining methods, and gathering necessary parts and tools. Scheduling, on the other hand, focuses on determining when the tasks will be executed. Planners should concentrate on planning and should neither schedule tasks nor execute them.

Develop Clear and Concise Job Plans

Job plans should be straightforward and include all necessary details, such as estimated time to complete tasks, required tools, and special materials. The instructions should be simple enough for even the least experienced technician to follow without needing additional clarification.

Encourage Comprehensive Feedback on Completed Tasks

Feedback is critical for refining maintenance plans. Technicians should provide detailed input through the work order system, describing what worked, what didn't, and any issues encountered. A simple status update like "complete" or "fixed" is insufficient for meaningful analysis.

Adapt Plans Based on Feedback

Planners should carefully review and incorporate technician feedback to improve work orders and processes. Addressing feedback not only enhances the effectiveness of maintenance plans but also fosters a culture where technicians feel valued and motivated to share insights.

Continuous Best Practices (3-4 Times Daily)

The following "Continuous Best Practices" should be carried out at least 3-4 times daily to maintain consistent visibility into the operations:

1. Regularly monitor the Spindle Displays throughout the day.
2. Check for equipment listed in Pending Maintenance for an extended period.
3. Identify equipment listed as Down for Maintenance for an excessive amount of time.
4. Look for Overdue PMs/Work Orders, which are highlighted in red font on the Spindle Display.
5. Conduct continuous Facility Page Checks to track equipment utilization.
6. Stay alert for active Alerts, including:
7. Pending Maintenance
8. Down for Maintenance
9. Overdue PMs
10. Overdue Work Orders
11. PMs taking excessive time

Daily Best Practices (15-30 Minutes Daily)

The following "Daily Best Practices" should be performed at least once daily and should take approximately 15-30 minutes to complete:

- **Login Compliance Checks (5-10 Minutes Daily):** Every morning, Maintenance Managers should review and clean up the previous day's data on the Analysis Page to ensure proper logging.
- **Overdue PMs/Work Orders Checks (5 Minutes Daily):** Review the CMMS Details Page every morning to identify overdue PMs or work orders.
- **Work Order Request Checks (5 Minutes Daily):** Check the CMS Details Page each morning to review work order requests that require Maintenance Manager approval or denial.

- **Approve/Deny Work Order Requests (5 Minutes Daily):** Approve or deny work order requests daily to ensure nothing is overlooked.
- **Create Work Order Requests (5 Minutes Daily):** As needed, create new work order requests to add to the Maintenance Schedule.
- **End of Day "End Shift" Checks (5 Minutes Daily):** Each afternoon, verify that all Maintenance Personnel have properly ended their shift in Spindle. If Maintenance Managers leave before some team members, designate a "Lead" to handle this responsibility.

Weekly Best Practices (1 Hour Weekly)

The following "Weekly Best Practices" should be performed at least once weekly and should take about 1 hour to complete:

- **Login Compliance Checks (10-15 Minutes Weekly):** Every Monday morning, review and clean up the previous week's data on the Analysis Page to ensure accurate logging.
- **General Maintenance Tasks (GMT) Review (10-15 Minutes Weekly):** Conduct a weekly GMT review on Monday mornings alongside Login Compliance Checks to confirm proper logging of GMT tasks.
- **Equipment Opportunity Checks (10-15 Minutes Weekly):** Perform weekly Equipment Opportunity Checks on Monday mornings, in conjunction with Login Compliance and GMT Reviews, to identify equipment with significant improvement potential.
- **Scheduled Maintenance Tasks Review (5 Minutes Weekly):** Review the Scheduled Maintenance Tasks (PMs and Work Orders) completed the previous week every Monday morning.
- **PM Efficiency Report Review (5-10 Minutes Weekly):** Analyze the efficiency of PMs and Work Orders completed last week every Monday morning.
- **Overdue PMs/Work Orders Checks (5 Minutes Weekly):** Every Friday, review overdue PMs or work orders for the current week to plan for the next week and avoid missed tasks.
- **Upcoming PMs/Work Orders Scheduling Review (10-15 Minutes Weekly):** Every Friday, examine the PMs or work orders scheduled for the upcoming week to plan tasks effectively and ensure proper scheduling and assignment.

Monthly Best Practices (1 Hour Monthly)

The following "Monthly Best Practices" should be performed at least once a month and require approximately 1 hour to complete:

- **Equipment Opportunity Checks (10-15 Minutes Monthly):** At the beginning of each month, conduct Equipment Opportunity Checks in conjunction with weekly checks to gain a broader perspective for analyzing equipment performance and identifying opportunities.
- **Scheduled Maintenance Tasks Review (20-30 Minutes Monthly):** At the start of the month, review the Scheduled Maintenance Tasks (PMs and Work

Orders) completed during the previous month and those scheduled for the current month. Adjust scheduling and worker assignments as needed.

- **PM Efficiency Report Review (10-15 Minutes Monthly):** At the beginning of each month, analyze the efficiency of PMs and Work Orders completed during the previous month.

MAINTENANCE AFTER 250 HOURS OF OPERATION

The maintenance after 250 hours of operation service includes the jobs which are necessary for tractor to tune-up, which are as below:

Repeating the Activities Carried out in 125hours Maintenance Schedule

Valve Clearance Adjustment

Proper valve clearance is very important because:

- Valves give longer life
- Increases fuel efficiency
- Starting becomes easy
- Maximum power generation

Figure: 17.1 View of Valve System in Tractors

Recommended valve clearances for tractor engine are given in table below:

Type of valve	Recommended valve clearance
Inlet valve	0.008″
Exhaust valve	0.010 to 0.016″ 0.004″ (1 thou) = 0.1 mm

Following steps may be followed to valve clearance adjustments:

Steps for valve clearance adjustment:

1. Check cylinder head bolts for tightness.
2. Slowly turn crankshaft until piston number 1, in the cylinder is at top dead centre (TDC) of compression stroke.
3. Check clearance with filler gauge between valve stem and rocker arm of both the valves.
4. If the clearance is not proper, loosen the adjusting screw nut and turn adjusting screw and check with filler gauge.

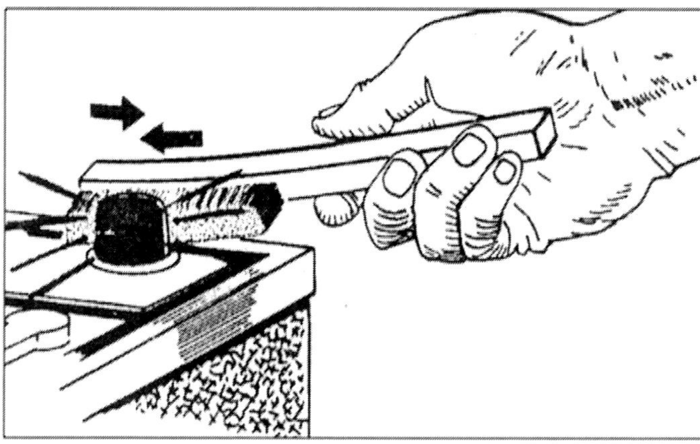

Figure: 17.2 View of removal of corrosion from battery terminal

5. Determine which cylinder fires next.
6. Turn crankshaft till the next cylinder in firing order is on compression.
7. Adjust valves, following the same procedure as on No. 1 cylinder.

Checking the Battery

- Disconnect cable and ground strap from the battery terminals.
- Clean cable clamps.
- Remove dust and corrosion particles from battery terminals.
- Brush soda and water mixture on top of the battery.
- Apply coating of petroleum jelly.
- Reconnect power cable and tighten properly.

Cleaning the Sediment Bowl and Fuel Filters

1. Remove the sediment bowl after closing the fuel supply.
2. Remove gasket, if any.
3. Remove the filter or strainer and wash it with diesel. Pass the compressed air through it, so that dust and dirt can be removed more effectively.
4. Clean sediment bowl also. Open fuel valve and check whether fuel is coming from tank is free from dirt or not, if not make a free flow:

Figure: 17.3 View of sediment bowl

5. Assemble the sediment bowl and reinstall it over to the tractor.
6. Change fuel filter elements as per recommendation of the manufacturer.

Adjust Tractor Brakes

- Check what provision is made for the brake adjustment.
- Jack up rear of the tractor until both the rear wheels clear the ground.
- Release brake interlocks and locks.
- Complete the preliminary arrangements for adjustment for first brake. Remove adjusting cover.
- Tighten adjusting screw, adjusting rod or adjusting nut. The brake is tightened by shortening the linkage between the brake pedal and the brake. It is tightened until there is 15-20 mm of free travel on the brake pedal between its released position and the point where you can feel the brake starting to take hold.

Figure: 17.4 View of Brake System

- Complete reassembling, or tightening of the locknuts to maintain adjustment.
- Adjust second brake in the same manner as explained above.
- Check to make certain that the brakes are equalized then lower tractor from the jacks.

Adjust Engine Clutch

Adjustment of foot Operated Clutches: To adjust the foot operated clutch proceed as follows:

- Determine from your operator's manual how much free travel in clutch pedal is needed. Recommendations vary from 15mm to more than 60mm.
- Check clutch pedal for free travel.

Figure: 17.5 View of Clutch

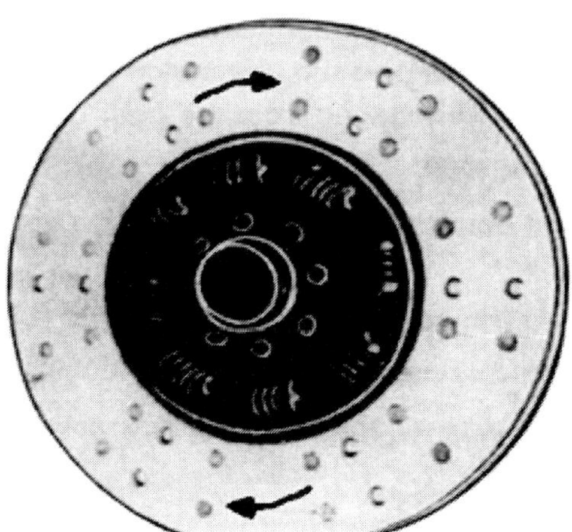

Figure: 17.6 View of disc type clutch

- Locate means provided for clutch adjustments.
- Adjust linkage until clutch pedal has sufficient free travel.
- Tighten lock nut so that it holds adjustment securely.

Adjustment of Hand-Operated Clutches: To adjust the hand-operated clutch proceeds as follows:

- Check clutch lever adjustment to determine whether adjustment is needed.
- Disengage the clutch.
- Place gear shift lever in neutral.
- Remove hand-hole cover on clutch housing
- Turn clutch by handle until locking mechanism is on side next to hand-hole.
- Do the adjustment
- Reassemble it.

18

Engine Overhauling and Assembling

INTRODUCTION

Tractors revolutionized agriculture by replacing animal-powered farming with machine-driven efficiency, significantly boosting productivity for millions of farmers globally. However, regular tractor engine maintenance is crucial to ensure smooth and efficient operation.

An engine overhaul is a thorough process designed to restore the optimal performance of an internal combustion engine. Over time, engines experience wear and tear due to constant use, exposure to different conditions, and aging, which can result in reduced performance, efficiency, and reliability. To mitigate these issues and prolong the engine's operational life, an overhaul becomes essential.

A tractor engine overhaul involves dismantling the engine to diagnose and replace defective components, followed by reassembly. This process aims to enhance the engine's performance, power, and overall efficiency, ensuring it meets the demands of agricultural work effectively.

WHAT IS AN ENGINE OVERHAUL?

An overhaul is an extensive procedure that includes disassembling, inspecting, cleaning, and repairing various engine components. This process aims to restore the engine to its original condition or enhance its performance. It is applicable to a wide range of engines, including those found in cars, airplanes, boats, and other machinery, with specific procedures varying based on the engine type and the extent of repair required.

An engine overhaul involves inspecting, cleaning, and either repairing or replacing internal components as necessary. Typically undertaken when an engine shows significant performance decline due to wear and tear, an overhaul can restore functionality and efficiency, offering a more cost-effective alternative to complete engine replacement.

From practical experience in the field, overhauls are often essential when engines become difficult to start, lose power, or consume excessive fuel. Addressing these issues promptly is crucial, as neglect could lead to further damage and eventual engine failure.

Diesel Engine Overhaul?

An overhaul of a diesel engine involves renewing or reconstructing all its key parts. Typically, this encompasses changing out the pistons, connecting rods, cylinder liners, bearings, and gaskets. In certain instances, it might also be necessary to replace the crankshaft and injectors.

TYPES OF ENGINE OVERHAUL

In the heavy machinery industry, engine overhauls are categorized into several types, depending on the severity of the engine's condition and the required repairs:

- **Minor Overhaul:** This type involves repairing minor components without disassembling the entire engine. Minor overhauls are typically done when the engine exhibits light issues, such as unusual noise or a slight loss of power. The process includes cleaning, oil changes, filter replacement, and inspection of key components like spark plugs or fuel injectors.
- **Top Overhaul:** This involves removing only the cylinder head, which is the upper part of the engine, while the engine block, the lower part, remains in the vehicle. In a top overhaul, repairs are focused on the upper part of the engine, including the cylinder head, valves, pistons, and combustion system. I've personally conducted a top overhaul on a genset engine that frequently overheated due to worn-out valves and a faulty cylinder head.
- **Major Overhaul (Full Overhaul):** This process entails taking out the entire engine and completely disassembling it. This is a complete repair that involves disassembling the entire engine, replacing or repairing critical components like the crankshaft, pistons, and engine block. A major overhaul is necessary when the engine has suffered significant damage or performance degradation.

THE BASIC STEPS OF ENGINE OVERHAUL

The repair process begins with dismantling the engine. This step entails taking out major components such as the cylinder head, crankcase, and pistons. After they are removed, each component is carefully inspected for any signs of wear or damage. Parts found damaged or excessively worn are replaced with new ones.

Subsequently, the engine undergoes a thorough cleaning to eliminate any accumulated dirt, debris, or other impurities. This cleaning process is often carried out with specialized tools such as high-pressure hoses and ultrasonic cleaning devices.

Once the engine is both cleaned and inspected, it is put back together. This reassembly may involve incorporating new or refurbished parts as required. Finally, the engine undergoes testing to confirm that it operates correctly and aligns with all the specifications set by the manufacturer.

ENGINE OVERHAUL PROCEDURE

While overhaul procedures vary based on the type of engine and the extent of the damage, the following are general steps commonly followed during engine overhauls based on my field experience:

- **Initial Diagnosis:** The first step is to conduct a thorough inspection of the engine to identify the main issues. This involves compression testing, checking for oil leaks, and analyzing engine noise.
- **Engine Disassembly:** After identifying the problems, the engine is disassembled to inspect its internal components. Key parts like the pistons, crankshaft, camshaft, and valves are removed and checked for wear or damage.
- **Component Cleaning:** Components that are not damaged are carefully cleaned to remove accumulated dirt and deposits. From my experience, keeping components clean is essential for ensuring efficient engine operation after the overhaul.
- **Replacing Damaged Parts:** Any components that are significantly worn out or damaged must be replaced with new ones. This could include pistons, bearings, piston rings, or the crankshaft if cracks or severe wear are found.
- **Component Reconditioning:** Some parts, such as the cylinder head or engine block, can be reconditioned to fix minor damage. Reconditioning involves surface grinding and pressure testing.
- **Reassembly:** Once all components have been repaired or replaced, the engine is carefully reassembled according to manufacturer specifications. This is a critical step to ensure all parts work harmoniously together.
- **Engine Testing:** After reassembly, the engine is started and tested to ensure its performance is restored. Testing includes measuring pressure, temperature, and overall engine performance.

WHAT CAUSES ENGINE OVERHAUL

Repair may be needed due to poor performance. If your engine isn't performing optimally, an overhaul can help identify and correct problems, which may include replacing or upgrading parts.

Poor Performance

Indicators of poor performance necessitating an overhaul include:

- **Diminished Power:** Less power output can signal wear or damage in engine components, which an overhaul can address.
- **Decreased Fuel Efficiency:** Increased fuel consumption may indicate inefficiency, diagnosable and repairable through an overhaul.
- **Unusual Engine Noises:** Knocking or pinging sounds suggest internal component issues, which can be identified and resolved through an overhaul.
- **Excessive Smoke:** High levels of smoke from the engine point to internal problems, correctable with an overhaul.

Damage

Damage is another common cause that requires repair. Whether due to an accident or the use of low-quality fuel, such damage often requires repair or replacement of the affected engine components.

Specific damage types that might lead to an overhaul include:

- **Physical Damage:** Accidents or other physical impacts can necessitate engine repairs or replacements, with an overhaul assessing and addressing the damage.
- **Water Damage:** Exposure to water, like in floods, may require engine repair or replacement. An overhaul can determine the damage's extent and needed repairs.
- **Corrosion:** Corroded engines may need repair or replacement, with an overhaul revealing the severity and required fixes.
- **Wear From Low-Quality Fuel:** Using substandard fuel can cause additional engine wear and damage, which an overhaul can help identify and rectify.

Neglecting Maintenance

Neglecting maintenance is also a major reason for repair. Regular maintenance, such as changing the oil, adhering to service intervals, and fixing problems promptly, is critical to the health and longevity of your engine. Failure to properly maintain a engine can lead to various problems.

COMMON FIELD INCIDENTS

During overhauls, various field incidents often arise. From my experience, here are some common occurrences:

- **Unexpected Oil Leaks:** One of the most common problems after an overhaul is oil leaks. These can be caused by improperly installed seals or gaskets. In some cases, the engine may need to be disassembled again to fix the issue.
- **Overheating After Overhaul:** I've encountered situations where the engine overheated after an overhaul, which was due to improper installation of cooling components. This highlights the importance of thorough inspection before reassembling the engine.
- **Lubrication System Failure:** A faulty lubrication system can cause components to wear out quickly after an overhaul. This often happens when oil channels are clogged or the oil pump is defective.

MAJOR OVERHAULING/DE-CARBONISATION

The process of dismantling, cleaning, and removing carbon deposits from the following parts.

- Piston
- Cylinder Liner
- Cylinder head

When Major Maintenance/Decarbonization is Needed

- Completion of running hours as per the planned maintenance program of OEM (Original equipment manufacturer).
- Accident/Breakdown.
- Before continuous surveys in which process of change of running parts like piston rings etc.
- There are standard procedures for maintaining these machines, irrespective of their design and manufacturing.

Key Considerations for Maintenance Planning

Makeshift arrangements for the emergency in the event of a Power failure

- Special Tools and tackles to be in proper operating condition.
- All measuring tools, gauging, and measuring devices should be in suitable working conditions with a reasonable calibration period.
- Check and ensure the availability of spares needed for maintenance.
- To ensure the working of Power Packs if available on board.
- Engineers/technicians involved in the care should know the procedure per the Operating manual.
- The maintenance team should have Knowledge of symbols and signs used to indicate the safety of machinery.

GETTING READY FOR AN OVERHAUL

Essential Preparations Before Maintenance

- Quarantine
- Tagging Out
- Inspection of special tools supplied by the Manufacturer.
- Knowledge of functioning and procedure for operating the devices.
- Referring to the correct tightening torque and hydraulic pressure values.
- Complete understanding of parts assembly.
- Preparation of safety regulations for safe working.

Before maintenance begins, it's crucial to quarantine the machine to prevent spills and accidents. This includes isolating both the prime mover and alternator for safety.

Steps to Isolate Electrical and Mechanical Systems

- Open the ACB (Air Circuit Breaker) to cut off the alternate and bus bar links.
- The power supply of the Pre-Lube Oil Pump.
- Oil supply to Engine
- Cooling water supply
- Lube oil

- Fuel oil supply
- Air supply
    - A checklist of all the isolated areas is to be prepared, and the same is to be referred to before reopening the systems.
    - During the isolating process/opening procedure, a slow action with close observation must be followed to ensure other equipment is not affected.
- **Add on Safety:** While the maintenance work is in progress, additional tagging is to be carried out to prevent any opening up of Valves or Electrical breakers resulting in an accident of a serious nature.

Checking Engine Manufacturer's Special Tools

- Injector and starting valve opening tool
- Spanners
- Cylinder head opening tool
- Cylinder Liner Removal Tools
- Cylinder Liner Holding Tool
- Cylinder head lifting tool
- Piston Lifting tool
- Bearing removal tool
- Cylinder Head and con rod nuts loosening and tightening hydraulic tool
- Valve grinding tool
- Valve seat grinding machine/tool
- Injector testing tool
- Rope and ceilings with proper size D shackles.
- Overhead cranes.
- Engine Barring arrangements
- Cylinder Liner Brushing tool
- Piston Ring removal tool
- Piston guide tool for inserting the piston.
- All the pressure and temperature calibration instruments.
- Electrical measuring instruments.

Assembly Guidelines to Follow

- The engineer in charge should know the assembly of parts and doubt the Instruction manual/Consult with the Senior Engineer.
- Store all the dismantled components in a separate box with the marking on significant parts. For each unit, a separate container is to be used.
- Mark all the bearings and con-rod bolts and wrap the threads with the cloth to avoid the threads' damage.

Perform the Following Tasks

Cylinder Head Maintenance Checklist

- Inspect cylinder heads for wear or damage.
- Examine valve spindles for proper functioning.
- Inspect Inlet and Exhaust valves, as well as Inlet valve guides, and record measurements.
- Replace corroded or worn-out valve guides with new ones.
- Inspect valve seats, rejecting excessively damaged ones and re-grinding accepted seats.
- Disassemble valve rotators for inspection, looking for deep markings on ball pocket surfaces.
- Inspect all Injector Sleeves and document observations.
- Verify and adjust safety valve operating pressure.
- Examine Indicator valves.
- Check cylinder head cooling water spaces.
- Lap cylinder head seating faces.
- Perform pressure testing on all cylinder heads.
- Conduct valve seat grinding for both old and new seats.
- Remove, clean, inspect, and reinstall Water guide jackets with new O rings.
- Properly tighten cylinder head nuts following OEM torque specifications.
- Ensure thorough maintenance for optimal performance.

CYLINDER LINER MAINTENANCE PROCEDURE:

Conduct inspection, bore measurement, and honing at the workshop following OEM guidelines for Cylinder Liner Inspection and Honing.

- Record size measurements in the provided forms.
- Perform lapping on the upper face of the cylinder liner.
- Replace all flame rings with excessively worn-out working surfaces.
- Follow these steps to ensure proper cylinder liner maintenance.

Piston Inspection and Overhauling Procedure

- Perform decarbonization on all pistons, ensuring thorough cleaning both inside and outside.
- Clean all piston and oil scraper ring grooves, then inspect and measure them for wear using the groove measuring tool provided by the OEM. Record the measurements.
- Replace all piston rings with new ones.
- Ensure that the piston ring positions are offset 180 degrees from each other, and the spring joint of the scraper ring is also fitted with a 180-degree offset to the joint during piston installation.

Figure: 18.1 Remove all Pistons for Inspection and Overhaul

- Verify the tightness of the piston plug at the specified torque.
- Follow these steps for proper piston maintenance and overhaul.

CENTRAL TOOLS 6434 SLEEVE HEIGHT AND COUNTER BORE GAUGE

Quickly measures the cylinder sleeve height using pointed contact; Counter-bore depth using flat end contact; Top dead center; Piston-to-deck height; Valve protrusion; Easy readings around the entire bore. This is a compact and very versatile gauge.

Connecting Rod Inspection Steps

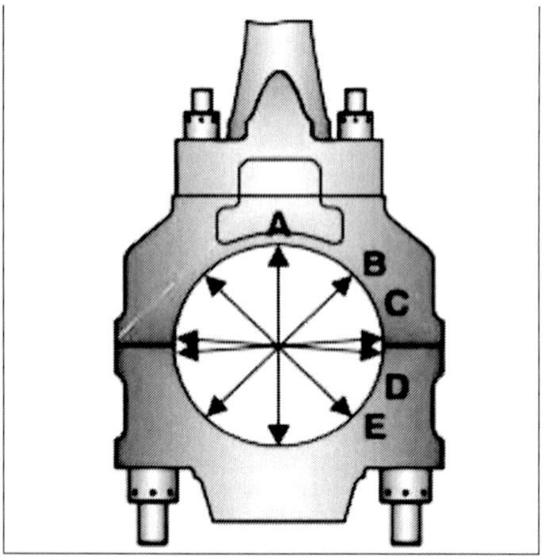

Figure: 18.2 Connecting Rod Inspection Steps

Inspect all connecting rod big-end bearing shells. Replace any with deep grooves or excessive wear beyond accepted norms for shell thickness and free spread.

- Examine the connecting rod's big end bore and document measurements.
- Check the small end bearings.

- Verify the clearance between the wrist pin and bush.
- Conduct an ultrasonic test on all piston pins to detect hairline cracks.

Gear Wheel and Camshaft Inspection

Inspect all gear wheel bolts and connections as per OEM (Original Equipment Manufacturer) guidelines to ensure secure and proper fastening.

- Replace camshaft bushes if they show signs of wear to avoid operational inefficiencies.
- Check the overall condition of the camshaft for any visible damage, cracks, or wear.
- Examine camshaft lubrication by observing the flow of lube oil from all bearings and spray pipe nozzles, ensuring the oil jet accurately targets the gear wheel's meshing point for proper lubrication.
- Verify the tightness of locking devices and fasteners to prevent looseness during operation.
- Inspect all pushrods for straightness and replace any that are bent to ensure proper functionality.
- Adjust the inlet and exhaust valve clearances following OEM-recommended specifications for optimal engine performance.
- Performing these inspections and adjustments is essential to maintaining the integrity, efficiency, and functionality of the gear wheels and camshaft.

Valve and Fuel Pump Mechanism Inspection

- Examine the Roller Guide for any marks.
- Check for deformation on the spherical stud.
- Inspect the surface of rollers for any marks or deformation.
- Verify the absence of free play between the rollers, the bush, and the pin.
- Inspect the Valve Gear, springs, and Pushrods.
- Examine all valve bridge springs, rocker arm brackets, and pushrods.
- Also, inspect the Roller Guide for fuel injection pumps.
- Ensure all roller guides are free of marks.
- Check the surface of rollers for defects and deformation.
- Confirm there is no free play between rollers, bush, and pin.
- Lastly, check the lubrication of all operating gears.

These inspections ensure the proper functioning of the inlet and exhaust valve mechanisms and fuel injection pumps.

GOVERNOR

- Overhaul the Governor.
- Change the Governor's oil.
- Crankshaft and Main Bearings Maintenance:

Figure: 18.3 Inspect all Crankpins and Polish all Main and Connecting Rod Pins with Glossy Paper.

Examine thrust bearings and replace them if wear exceeds prescribed limits.

- Re-tighten counterweights to the specified torque.
- Disassemble all main bearings and polish all crankpins.
- Check for runout.
- Tighten the main and guide bearing caps to the prescribed torque.
- Inspect and replace any main bearings with abnormal wear.

Follow these steps for proper maintenance of the crankshaft and main bearings.

Turbocharger and Charge Air Cooler Maintenance

- Remove the Turbocharger (TC) cartridge.
- Replace the nozzle ring.
- Dismantle the cartridge, inspect the bush and labyrinth ring, and replace if necessary.
- Re-assemble the cartridge.
- Check clearances following the procedure.
- Clean the Charge Air Cooler (CAC).
- Perform a pressure test on the Charge Air Cooler.

These steps ensure proper maintenance of the turbocharger and charge air cooler.

Engine Frame and Bed Plate Maintenance

- Re-tighten the holding-down bolts.
- Re-tighten the bolts connecting the engine and base frame.
- Conduct a functional test of the safety cover.

- Inspect the AVM pad adjust screw clearance, and make adjustments if necessary.

These steps ensure the proper maintenance of the engine frame and bed plate.

Fuel Oil System Maintenance

- Calibrate the fuel injection valves.
- Inspect all high-pressure fuel pipes and replace "O" rings.
- Examine the FIP roller and measure nominal height.
- Perform an overhaul of all fuel injection pumps, replacing the barrel, plunger, and 'O' ring kit as needed.

These steps are essential for the proper maintenance of the fuel oil system.

Lube Oil System Maintenance

- Remove the engine-driven lube oil pump for inspection.
- Dismantle the entire pump and replace the bushes.
- Apply the correct torque to tighten the self-locking nut.
- Drain old oil from the engine and refill with new oil.
- Overhaul the pre-lube oil pump.
- Replace the lube oil filter cartridge.
- Clean and replace the Glacier bypass filter kit.
- Perform a chemical circulation cleaning of the lube oil Plate Heat Exchanger (PHE).

These steps are crucial for maintaining the lube oil system.

Water Pump and Thermostatic Valve Maintenance:

- Remove the LT (Low Temperature) and HT (High Temperature) water pumps from the engine.
- Perform a comprehensive overhaul of the pumps.
- Replace the rotating sealing and both ball bearings of the HT Water Pump.
- Replace the rotating sealing and both ball bearings of the LT Water Pump.
- Tighten the self-locking nuts of both pumps to the specified torque.
- Overhaul the HT water thermostatic valve.
- Clean the LT/HT Plate Heat Exchanger (PHE).

These steps are essential for maintaining the water pumps and thermostatic valve in optimal condition.

Mechanical Overspeed Device Maintenance:

- Perform an overhaul of the mechanical overspeed device.
- This step ensures the proper maintenance of the mechanical overspeed safety mechanism.

Miscellaneous Procedures

These are crucial steps in the Diesel Generator Overhauling Procedure:

- Record the Fuel pump index at Governor Index 0.
- Verify the integrity of the compressed air system.
- Inspect the fuel oil system and nozzle cooling system.
- Perform functional tests on safety shutdown switches and analog switches.
- Calibrate all safety switches.
- Consider replacing the alternator bearing if necessary.
- After completing maintenance, initiate the D.G. Set and follow the sequence for all moving parts per the standard Diesel Generator Overhauling Procedure.
- Conduct load trials of the D.G. Set according to the OEM protocol.
- Record the set's performance during stage load sequences, ranging from 25% to the maximum available load, as specified in the manual.
- Continuously monitor parameters and log them in the logbook for reference.
- These steps ensure a comprehensive and successful Diesel Generator Overhauling Procedure.

IMPORTANCE OF POST-OVERHAUL MAINTENANCE

After an engine overhaul, regular maintenance is key to keeping the engine in optimal condition. From my experience, many issues can be prevented with routine maintenance, such as regular oil changes, cooling system checks, and engine pressure monitoring. SSC Works also offers post-overhaul inspection and maintenance services to ensure your engine continues to perform at its best.

Tractor Engine Maintenance

Proper tractor engine maintenance is essential for ensuring long-term performance and longevity. Here are the foundational steps to follow for optimal engine health:

- **Regular Oil Changes:** Changing the oil regularly is crucial for maintaining engine performance. Engine oil reduces friction, prevents wear, and extends the life of the engine. Adhere to the manufacturer's recommended intervals for oil changes, which could depend on operational hours or time duration.
- **Air Filter Control:** The air filter ensures the engine receives clean, oxygen-rich air. A clogged filter can lead to combustion issues and reduced performance. Regularly inspect, clean, or replace the air filter, especially if operating in dusty environments.
- **Fuel Filter Maintenance:** The fuel filter ensures clean fuel reaches the engine. A dirty filter can cause engine damage, so check and replace it regularly according to the manufacturer's guidelines.

- Cooling System Inspection: The cooling system helps regulate engine temperature. Regularly check the coolant level and top it up to avoid overheating, particularly during prolonged use in hot weather or heavy-duty tasks.
- Cylinder Head Gasket and Piston Ring Evaluation: Inspect the cylinder head gasket and piston rings regularly to prevent serious engine damage. Address any leaks or damages promptly, as they can cause a significant decrease in engine efficiency.

Tips for Enhancing Tractor Engine Performance

To enhance engine performance and improve work efficiency, consider these advanced tips:

- **Engine Modifications:** Work with professionals to make suitable modifications, such as adjusting engine parameters, upgrading components, or installing performance kits. These modifications can increase engine power but must comply with local regulations and safety standards.
- **High-Quality Fuel Usage:** Use high-quality diesel fuel with the correct cetane rating to ensure efficient engine operation. Purchase fuel from reputable sources to prevent contamination.
- **Turbocharger or Supercharger Addition:** Adding a turbocharger or supercharger increases engine power by providing more air and fuel. Ensure professional installation to guarantee compatibility with your tractor's engine.
- **Upgrading Exhaust Systems and Filters:** Enhance engine efficiency and reduce emissions by upgrading the exhaust system and installing advanced filters. This will improve performance and contribute to environmental sustainability.
- **Regular Inspections and Adjustments:** Periodically inspect your tractor for potential issues and necessary adjustments. A well-maintained engine, with clean components and securely connected parts, operates more efficiently and reliably.
- **Tyre Condition:** The condition of the tires directly affects tractor performance. Maintain proper tire inflation, check for wear, and ensure tread depth is sufficient for optimum traction and control.
- **Engine Temperature Monitoring:** Use thermometers or monitoring systems to track engine temperature. This allows for early detection of overheating and timely corrective actions.
- **Load Balancing:** Avoid overloading the engine and respect the tractor's maximum carrying capacity. Proper load distribution ensures safety and prolongs the engine's lifespan.
- **Professional Maintenance:** Have your tractor regularly serviced by experienced professionals who specialize in tractor engines. They can identify and address potential problems early, ensuring peak performance.

Tractor engines are vital assets in agriculture, construction, and transportation.

Prioritizing both maintenance and performance enhancement is essential for efficiency and the longevity of the equipment. A well-maintained tractor engine is the key to achieving work success and operational efficiency.

TRACTOR ENGINE SPARE PARTS MAINTENANCE

To extend the lifespan of tractor engine spare parts and ensure optimal engine performance, regular maintenance is essential. While the maintenance program may vary depending on the model and usage, the general steps include:

- **Changing Oil and Oil Filter:** Engine oil lubricates and cools engine parts. Over time, the oil becomes dirty, and its viscosity decreases. It's crucial to change both the engine oil and oil filter according to the manufacturer's recommended intervals to maintain efficiency and protect engine components.
- **Replacing Air Filter:** The air filter ensures that clean air enters the engine. A clogged or dirty air filter can decrease engine performance and increase fuel consumption. Regular replacement at the manufacturer's suggested intervals helps maintain engine power and fuel efficiency.
- **Checking the Cooling System:** The cooling system prevents overheating, ensuring the engine runs at an optimal temperature. Regular checks of the coolant level and condition are essential, and the coolant should be replaced when necessary to maintain proper engine cooling.
- **Inspecting Batteries:** Batteries are essential for starting the engine. It's important to inspect the batteries regularly, ensuring they are charged and in good condition. Replace them when necessary to avoid starting problems.
- **Inspecting Spark Plugs:** Spark plugs ignite the fuel in the engine. Over time, they can wear out or become dirty, affecting engine performance. Regular inspection and cleaning, or replacing spark plugs as needed, ensures smooth engine operation.
- **Checking Belts:** Belts are responsible for transmitting power between various engine components. Regular inspection of belts for wear and tear is crucial. Replace any damaged or worn belts to prevent engine failure and maintain power transmission efficiency.

By adhering to these maintenance steps, you can ensure the smooth operation and longevity of your tractor engine, reducing the likelihood of breakdowns and costly repairs.

Tractor Engine Spare Parts Repair

In case of tractor engine spare parts failure, it's important to determine the cause of the failure and perform the necessary repairs. To determine the cause of the failure, listening to the engine's sound, checking for smoke and leaks, and conducting necessary tests are required. Once the cause of the failure is identified, the faulty part should be repaired or replaced.

Things to Consider

During tractor engine spare parts maintenance and repair, consider the following:

- Always prioritize your safety. Take necessary safety precautions when working on the tractor.
- Use the correct tools and spare parts. Using incorrect tools or spare parts can damage the engine or cause injury.
- Be patient and work carefully. Engine repair can be complex, so staying calm and working carefully is important.
- Seek help from an expert if needed. If you're not comfortable with engine repair or struggling, don't hesitate to seek help from an expert.
- Ensure you use the manufacturer-recommended brand and viscosity of oil when changing oil and oil filters.
- When replacing the air filter, clean the filter box to protect the filter from dirt and dust.
- When checking the cooling system, ensure the coolant level is above the minimum level.
- When inspecting batteries, ensure the battery terminals are clean and free of rust.
- When checking spark plugs, ensure the electrodes of the spark plugs are not worn or damaged.
- When checking belts, ensure the belts are tight and free of cracks.
- Prepare necessary spare parts and tools in advance when performing engine repair.
- Keep the work area clean and organized when performing engine repairs.
- Maintenance and repair of tractor engine spare parts are crucial aspects when working in the field. By following the tips mentioned in our article and taking necessary precautions, you can extend the life of your engine and ensure its smooth operation.
- Remember, improving your knowledge and skills in tractor maintenance and repair is always beneficial, allowing you to intervene more quickly and easily in emergencies.

TRACTOR ENGINE MAINTENANCE BEST PRACTICES

There are a number of things you can do to maintain your tractor engine, including changing the oil and filter, checking the air filter, and inspecting the engine for leaks and damage.

Changing the Oil and Filter

The oil and filter are two of the most important parts of your tractor engine. The oil helps to lubricate the engine and keep it running smoothly, while the filter helps to remove dirt and debris from the oil.

Always remember to change the oil and filter regularly, according to the manufacturer's recommendations. Each time you change them, you are saving your engine from damage and ensure it remains clean longer.

Checking the Air Filter in Regular Tractor Engine Maintenance

The air filter helps to clean the air that is entering the engine. A dirty air filter can restrict airflow to the engine, which can cause it to overheat or run inefficiently.

Check the air filter every now and then, and replace it as needed. And never underestimate of your engine's need to breathe.

Inspecting the Engine

It is imperative to Inspecting your tractor engine regularly for leaks and damage is imperative to keep it functional at high efficiency. When inspecting your engine, look for:

1. Signs of leaks
2. Cracks
3. Other damage

If you spot any signs of damage, have a professional repair the engine immediately. This can and will cost you some money, but the amount cannot be compared to the amount you would pay if you neglected the issue.

Using the right Diesel Tractor Oil

Using the right diesel tractor oil is essential for keeping your tractor's engine running smoothly. The tractor oil you choose should be designed for use in diesel engines and should meet the manufacturer's specifications.

There are a number of different types of diesel tractor oils available, so it is important to choose the right one for your needs.

19

Implement Hitching and Detaching

INTRODUCTION

Hitching and detaching implements from a tractor is a crucial aspect of tractor operation, whether in agricultural, industrial, or construction work. This process involves attaching various implements, such as plows, harrows, or trailers, to the tractor's hitch system, ensuring proper functionality during operation, and safely detaching them once the task is completed. Mastering the correct procedures for hitching and detaching is vital for safety, operational efficiency, and preserving the tractor and implement lifespan.

The tractor hitch system plays an essential role in connecting the prime mover (tractor) to the implement, creating a cohesive working unit. Both the tractor and the implement rely on each other to perform, much like the two wheels of a cart. A "hitch" is defined as a single articulated point or a combination of points and links that allow the tractor to exert tractive effort in the form of pushing or pulling to counteract the draft force of the implement. The hitch point, whether virtual or real, represents the specific location on the tractor through which the "line of push" must pass to ensure effective operation. Proper understanding of this system ensures smooth, safe, and efficient work.

HITCHING IMPLEMENTS TO THE TRACTOR

Hitching implements to a tractor is an essential process, and the method can vary depending on the type of implement being used. The most common method involves using two hydraulic arms and a third high point to attach and level the implement. The hydraulic arms serve as the main tie-down, allowing the implement to be raised or lowered, while the third point is connected through a double-nut bar, which can adjust the length of the tie-down. For larger tractors or implements, this third point can also be hydraulic for easier adjustment.

An alternative coupling method is the cup and pin system, commonly used for trailers, balers, or tanks, where a pin is driven through the tractor's cup and the implement's drawbar. Another method, the power take-off (PTO), is used for implements requiring rotational force, such as milling machines or brush cutters. The PTO is a bar that uses the tractor's engine revolutions, with a cardan system to transmit the rotational motion to the implement.

Regardless of the type of hitching system used, the process involves carefully backing the tractor toward the implement, using the hydraulic controls to position the arms close to the implement's hitch point. The implement should be secured with a jack or kickstand to prevent movement, and the engine should be turned off before attaching it to the tractor. It's crucial to ensure the tractor is parked on firm ground and is stable before hitching.

Improper hitching can lead to serious accidents, including rollovers, which can result in injury or death. NIOSH data suggests that improper hitching is responsible for a significant portion of rear rollover incidents. To reduce risks, NIOSH recommends:

- Proper training for tractor operators.
- Using farm equipment with rollover protective structures and always wearing a safety belt.
- Carefully selecting the hitching point and never attaching a load directly to the axle.
- Never using a two- or three-point hitch in place of a drawbar and only using the drawbar for single-point attachments.
- Ensuring the tow chain is sufficiently strong and long enough for a safe stopping distance.
- When using a tractor to free an embedded vehicle, hitching vehicles front-to-front and driving the towing tractor in reverse to reduce rollover risk.

HITCH SYSTEMS AND HITCHING TILLAGE IMPLEMENTS:

The force relation in hitching pull-type implements involves analyzing various forces acting on the tractor and implement during operation. These forces include soil force components (L, S, and V), the implement's gravitational force (W), and the parasitic soil forces (Q). The forces acting on the hitch system can be categorized as independent and dependent variables.

Independent Force Variables:

1. **L (Longitudinal soil force):** The force acting along the direction of travel.
2. **S (Side force):** The force exerted by the soil perpendicular to the direction of travel.
3. **V (Vertical soil force):** The force exerted vertically by the soil on the implement.
4. **W (Weight of the implement):** The gravitational force acting on the implement due to its weight.

Dependent Force Variables

- **Q (Parasitic soil forces):** Forces that oppose the movement of the implement through the soil, which depend on the hitch arrangement.

- **P (Pull force):** The force exerted by the tractor to move the implement, which also depends on the hitching system.

In horizontal hitching, the forces include:

- **R:** The horizontal component of the pull force.
- **Q:** The parasitic soil force opposing the tractor's movement.
- **P:** The actual pull force required by the tractor to tow the implement.
- **W:** The gravitational force on the implement, which is typically vertical but can affect the horizontal forces.

In vertical hitching, the components of the forces are analyzed in the vertical plane, considering how vertical forces influence the depth and weight distribution of the implement in the soil.

For pull-type implements with adjustable pull members, the primary objective is to establish the location and magnitude of the resultant parasitic support force (Qh or Qv) and pull (Ph or Pv). These forces are influenced by the hitch design and the method of controlling the implement's depth and movement through the soil. Proper hitching ensures the forces are balanced to avoid excessive stress on the tractor or implement and to maintain effective operation.

For mounted or semi-mounted implements, the force relation is largely determined by the design of the hitch linkage, implement configuration, and depth control mechanisms, rather than by the hitch adjustments themselves. The forces here depend on the structural design and control system for maintaining proper depth and stability during operation.

HITCH SYSTEM COMPONENTS

Three point hitch system is the only hitch system now available for all implements except trailers and combines. Thus, information will be restricted to three point hitch system. Three types of hitching possible with three point hitch system are free float (towed), semi-mounting, and fully mounting. The components of a three point hitch system and their functions along with numbers required are given in Table.

Table 19.1 Major functions of various components of a hitch system

Component		No. required	Functions
1.	Upper Link	1	Basic component of hitch system. Connects implement to tractor. Transfers hitching forces of tractor to implement and of implement to tractor
2.	Lower Link	2	-do-
3.	Lateral Limiters	2	Limits the travel of implement in horizontal plane thereby inhibits the horizontal sway.

Component		No. required	Functions
4.	Lift Link	2	Lifts the implements and supports the implement in transport position. It provides lateral stability and lateral levelling to the implement.
5.	Lift Arm	2	It provides lateral stability. It lifts implement by giving needed force and motion to lift link and supports implement in transport position.
6.	Rocking Arm	1	It connects cylinder and rockshaft. It converts linear motion of cylinder into rotation.
7.	Rockshaft	1	It transmits the rotating motion of rocking arm to lift arm.
8.	Cylinder	1	It is source of power for lifting implement. It gives linear motion to rocking arm.

Vertical Hitching of Pull Type Implement

Types of Vertical Hitching Systems

Pull: type tillage implements fall into one of the following three categories:-

- Implements with hinged pull members that have support wheels or support runners to gage the depth. The pull members act as a free link in the vertical plane, e.g., M.B plow, disc plow and drag type spring tooth harrow, etc.
- Implements with hinged pull members that do not have gage wheels or runners. The only support is through soil-working units and parasitic forces cannot be separated from useful soil forces, e.g., disc harrows without wheels, spike tooth harrows and tandem-gang rotary hoes.
- Single-axle implements with rigid pull members, e.g., field cultivators, chisels, sub-soiler.

HITCHES OF MOUNTED IMPLEMENTS

Design Considerations and Types of Linkages

Generally two types of hitch linkages are used on tractor rear mounted three point converging-link types. The rear mounted three point parallel link type hitches for front mounted cultivators are not common in India. Single axis hitches are replaced by three point hitches. Any of these types can be operated with hitch members acting as free link in vertical plane or restrained links.

- Factors considered in designing or evaluating a system for mounting tillage implements on rear of tractors:-
- Ease of attachment and adjustment, versatility and safety.
- Standardization to permit inter- changeability.
- Uniformity of tillage depth as the tractor passes over ground surface irregularities.
- Ability to obtain penetration of implement under adverse conditions, particularly with implements such as disk harrows and disk plows.
- Rapidity with which tools such as plows as listers enter the ground.
- Trailing characteristics of implement around contours and on side hills.
- Effect of implement upon tractive ability of tractor (vertical load transfer from front wheels to rear wheels).
- Effect of raised implement upon the transport stability of the tractor.

Three Point Hitches

Hydraulic control systems were introduced on tractors in late 1930s. No. of hitching arrangements for rear mounted implements were developed. Most common among these is three-point hitch system. The two lower links cover toward the front and are free to swing laterally within limits. The lower links can also be locked. So, they are laterally rigid, which is desirable in some non-tillage applications. ASAE-SAE standards for 3-point hitches specify all dimensions related to 3 connecting points between implements and tractor and minimum limits for lift height, Lateral leveling adjustments, side sway and minimum lift force to be available at hitch pins. Link lengths and amount of horizontal and vertical convergence are not specified. Maximum drawbar power available for different ranges of categories are:-

- Category I - 15 to 35 kW (20 to 45 hp)
- Category II - 30 to 75 kV (40 to 100 hp)
- Category III - 60 to 168 kW (80 to 225 hp)
- Category IV - 135 to 300 kW (180 to 400 hp)

Quick attaching couplers for three point hitches have been developed to facilitate the attachment of implements that are too heavy to be nudged into position by one man. These have been standardized to fit on the standard three-point linkage system. Quick attaching couplers allow the operation to couple or uncouple an implement without leaving the tractor seat thus contributing to both convenience and safety.

Automatic Draft Control

This system is a type of restrained link mechanism designed to automatically adjust the implement's depth to maintain a consistent preselected draft. When soil resistance is uniform, depth fluctuations caused by uneven terrain are minimized with automatic draft control compared to automatic position control. However, in the field, variations in soil resistance can lead to depth changes, regardless of whether the ground is smooth or uneven. On smooth fields, the automatic draft control system ensures that the average draft remains within the tractor's available power or tractive capacity.

If an overload condition arises, the system attempts to lift the implement, countering gravity, inertia, and any downward soil force components. This lifting action at the rear of the tractor generates a counterbalancing effect on the front wheels, temporarily transferring vertical load from both the implement and the front wheels to the drive wheels, thereby reducing wheel slippage. If the excessive soil resistance continues over a significant distance, the implement's depth is gradually decreased to maintain optimal operation.

Draft Sensing

Earlier upper link draft sensing was employed on all automatic draft control systems and it is still in use on small and medium size tractors. With mounted implements the upper link is always in compression when implement is in ground.

HORIZONTAL EFFECTS OF HITCHING

A key feature of a mounted implement is its ability to maintain a uniform width of cut when operating along contours without impacting the tractor's steering. If the rear-mounted implement does not allow for lateral movement relative to the tractor, it will cut outside while following contours, and poor steering response will occur due to the side forces generated by the implement. A lateral swinging hitch facilitates easier steering but may cause the implement to cut soil to the inside on a corner. A mounted implement on a three-point hitch, as shown in the figure, has lower links that converge towards the front and are free to swing laterally.

Parallel Link Hitches

Mounted row crop cultivators commonly use this system because raising or lowering the gang or toolbar adjusts the depth of all shovels equally without altering the pitch of the shovels. The lateral rigidity of the hitch is crucial to allow for close cultivation near plants. In the case of restrained link operation, the depth of cultivation is controlled through the listing mechanism. Free-link operation with gauge wheels is typically used on wide toolbars and multiple-gang cultivators. The virtual hitch point for free-link operation is located at the intersection of parallel links or at infinity. The vector Pv is parallel to the links, and the magnitude of Qv can be adjusted by altering the slope of the links. Moving the gauge wheel forward or backward does not affect the magnitude of Qv.

Load transfer Systems for pull –type Implements

Load transfer with pull-type implements plays a significant role in enhancing the tractive ability of larger tractors. To achieve this, an arrangement is needed to apply a moment to the implement's pull member, which in turn works to lift the rear end of the pull member and the front of the implement. The pull member must be sufficiently strong to withstand these forces. The vertical load from the implement and the tractor's front wheels is transferred to the tractor's drive wheels, contributing to the reduction of the implement's depth. This load transfer is influenced by several factors, including soil conditions, the mass and type of implement, the method used for controlling or gaging the depth, as well as the tractor's front-end stability and steering response. To facilitate this load transfer, modifications or additions to the implement's pull member are required. Some systems pull from the regular tractor drawbar and apply a consistent lifting force. This lifting force is achieved through an adjustable hydraulic pressure applied to the rockshaft lift cylinder. This hydraulic system provides constant force and can be adjusted as necessary to ensure the implement maintains the desired depth while also improving the tractor's traction and efficiency.

SAFETY MEASURES WHEN HITCHING IMPLEMENTS

Ensure that the anchor points are adjusted to match the specific force and load of the implement, ensuring a precise fit for each hole. It's also beneficial to have a pin attached to a chain, ensuring a secure and accurate installation. Three-point hitches function by attaching all three shear bolts to the tractor implement, eliminating the need for chains or similar tools for attachment.

Regarding power take-offs, they must be safeguarded to prevent anyone from getting caught, and care must be taken to avoid electrical cables being snagged or crushed by the hydraulic arms. Finally, ensure that the hitch attachment is raised off the ground before moving the tractor to avoid potential hazards.

Properly coupling agricultural implements is crucial when preparing for agricultural work, along with using the right machinery or spare parts. In agriculture, transporting large quantities of goods is common, so it's increasingly important to incorporate tractor hitches that enhance load capacity at the rear of the vehicle or improve overall performance. Understanding the different hitch options available helps in selecting the one best suited to each specific need.

20

Safety Rules

INTRODUCTION

Tractors and power tillers have revolutionized modern agriculture, enabling farmers to perform heavy-duty tasks with greater efficiency and precision. These machines have become indispensable tools for planting, harvesting, and tilling, contributing significantly to increased productivity in the agricultural sector. However, despite their immense benefits, tractors and power tillers present substantial safety risks if not operated correctly. The complexity of these machines and the environment in which they operate—often rugged and challenging—can lead to dangerous situations that pose a risk not only to the operator but also to those working nearby.

The safety concerns associated with tractors and power tillers stem from various factors, including the large size and weight of these machines, their powerful engines, and the high-speed operations they are capable of. The potential hazards include rollovers, being caught in moving parts, and accidents involving attachments or implements. Rollover accidents are particularly common and represent one of the most serious risks, with fatal outcomes often resulting from improper handling of the tractor, poor terrain conditions, or lack of proper safety features.

In addition to mechanical risks, there are environmental factors such as visibility issues, rough terrain, and unpredictable weather conditions that further complicate the safe operation of tractors and power tillers. It is critical to remember that even experienced operators can fall victim to accidents if they become complacent or ignore basic safety protocols.

This chapter on Safety Rules aims to equip tractor and power tiller operators with the essential knowledge and practices to ensure safe operation. By adhering to established safety guidelines, farmers can protect themselves, their workers, and those around them from accidents and injuries. Proper training, the use of protective equipment, regular maintenance, and careful attention to the operating environment are all vital components of a comprehensive safety strategy. We will explore various aspects of tractor and power tiller safety, including the importance of pre-operation inspections, safe driving techniques, personal protective equipment (PPE), and emergency preparedness. Furthermore, we will discuss the significance of maintaining the machinery in good working condition to prevent malfunctions and breakdowns that could lead to accidents. Ultimately, this chapter underscores the importance of cultivating a safety-first mindset among all users of agricultural machinery. Safety is not just a set of rules to follow but a crucial aspect of preserving life and ensuring the sustainability of agricultural work.

GENERAL SAFETY PRECAUTIONS

Before you drive the tractor for the first time, read the manufacturer's operating instructions and complete the necessary specialist training. It if it's been a while since you've driven the tractor, refresh your knowledge.

Ensure a rollover protective structure (ROPS) is fitted in accordance with s. 216 Roll-over protection on tractors of the Work Health and Safety Regulation 2011 and in compliance with the design and testing requirements in AS1636.1-1996 Tractors – Roll-over protective structures – Criteria and tests or an equivalent standard

- Wear comfortable, well-fitting clothing and boots
- Wear hearing protection on tractors without cabins
- Wear a seat belt if one is fitted
- Keep children away from tractors and machinery
- Keep all guards in place, including power take off and master shield guards

Do not carry passengers unless there is a designated seat and seat belt within the ROPS.

Hazards Associated with Tractors and Power Tillers

- **Roll-Over Hazard:** Tractors can overturn, especially on slopes or when performing sharp turns. Always use a Roll-Over Protective Structure (ROPS) and seatbelt to reduce injury risks.
- **Power Take-Off (PTO) Hazards:** PTO shafts can cause entanglement injuries. Keep a safe distance, and always disengage PTO when not in use.
- **Moving Parts:** Tractors have numerous moving parts like the wheels, axles, and tilling equipment that can cause severe injuries if they come into contact with operators.
- **Falling from the Tractor:** Always maintain three points of contact when getting on or off the machine. Use handrails, and avoid jumping off.
- **Exhaust Fumes:** Tractors can emit carbon monoxide. Ensure that exhaust systems are properly maintained and avoid using machinery in enclosed spaces.

- **Noise Exposure:** Long exposure to loud engine sounds can lead to hearing damage. Always wear ear protection when operating machinery.

Workplace Hazards

- **Uneven Terrain:** Tractors and tillers are used on fields, which may have uneven ground, rocks, or slopes. Always drive carefully, especially on steep inclines.
- **Improper Loading & Unloading:** Ensure proper loading and unloading techniques are followed to prevent tipping or loss of balance.
- **Crowded Work Environment:** Ensure that workers or bystanders maintain a safe distance from machinery in operation. Always use warning signs or flags.
- **Electrical Hazards:** Stay away from overhead power lines when operating equipment, especially when raising the loader or other components.

Tractor Safety Precautions

- **Proper Seating & Harnessing:** Always fasten the seatbelt when operating the tractor to prevent injury in case of a rollover.
- **Safe Use of Attachments:** Ensure that all implements (plows, harrows, tillers, etc.) are securely attached and operate within the recommended limits.
- **Speed Control:** Maintain a safe speed for the conditions of the field and terrain. Avoid excessive speeds that might lead to tipping or loss of control.
- **Braking System:** Always check that the braking system is functioning properly before use, especially when operating on inclined surfaces.
- **Engage PTO Carefully:** Never attempt to repair or adjust the PTO system when the engine is running.
- **Avoid Sharp Turns:** Make smooth turns and avoid turning too sharply while working on slopes to prevent rollovers.

Power Tiller Safety Precautions

- **Ensure Stable Placement:** Make sure the power tiller is stable on level ground before starting.
- **Handle with Care:** Power tillers are typically lighter and smaller, but they can still be dangerous. Always control the tiller with both hands, keeping a firm grip on the handles.
- **Guarding Against Entanglement:** The rotating blades of a tiller can entangle clothing, shoes, or long hair. Keep these away from the working area.
- **After Use Care:** Turn off the engine and allow the machine to cool down before cleaning or servicing.
- **Check for Loose Parts:** Ensure all parts, including blades and handlebars, are secure before starting the tiller.

Emergency Procedures

- **Know the Emergency Shutoff:** Learn where the emergency shutoff is located on the machine and how to use it in case of malfunction.

- **First Aid Knowledge:** Have basic first aid skills, especially in treating cuts, burns, and injuries caused by machinery.
- **Reporting Incidents:** In the case of an accident or near-miss, report it immediately, and take corrective actions to avoid future risks.

SAFE STARTING AND STOPPING

Once you're in the driving position and before you start the tractor, adjust the seat so that all controls can be operated comfortably and safely.

Then:

- Start the tractor from the driving position and not from the ground
- Do not step down from a moving tractor
- Ensure the park brake is on and operating effectively before exiting
- Do not park a tractor on a steep slope
- Remove the starting key when the tractor is not in use.

Safe Operation

While tractors are designed to operate in a range of different settings and on varied terrain, the unexpected can happen. To ensure the continued safe operation of your tractor, drive at speeds slow enough to retain control in any circumstance.

Other things you can do to stay safe include:

- Reversing out when stuck in soft ground and getting help if unsuccessful
- Watching for ditches, embankments, and depressions—crumbling and slippery banks can cause rollovers
- Reducing your speed before turning or applying brakes
- Reversing up steep slopes rather than driving forward (ensuring no trailing implements are attached)
- Using as wide a wheel track as possible when working on hillsides and sloping ground
- Descending slopes cautiously, keeping the tractor in low gear and allowing the motor compression to act as a brake
- avoid floodwater
- Following the manufacturer's instructions for adding ballast to improve tractor stability.

Safe Hitching

Always fit attachments to your tractor according to the manufacturer's instructions.

In addition:

- When attaching equipment, only use the mounting points or the draw bar provided by the manufacturer, do not use improvised methods
- Do not alter, modify, or raise the height of the draw bar outside of adjustments made by the manufacturer

- Never hitch above the centreline of the rear axle, around the axle housing, or to the top link pin
- When a power implement is attached, ensure all guards are in place before operating
- Do not attempt to adjust or work on implements while they are in motion
- Do not use or attach implements unless the power shaft or power take-off shaft is guarded.

Strategies for Mitigating Tractor Accident Risks

Be Physically and Mentally Fit

There are number of human factors involved in fatal tractor-related accidents. Certain factors like poor judgment, poor attitude, insufficient knowledge or training, fatigue, haste, stress, depression, intoxication, or showing off can cause a fatal tractor overturn accident.

Each operator must be physically and mentally fit and fine when operating a tractor. An operator who is sleepy, tired or not feeling well may not be able to react in time to avoid an accident. Your tractor does what you make it do.

Be Properly Trained

A person who does not have sense on how to operate a tractor with safety in potentially hazardous situations can be injured or killed by exercising poor judgment. Make sure all persons permitted to operate tractors have been thoroughly trained. A good place to start training is with the operator's manual. Review the operator's manual, if possible with the tractor in front of you.

Be Familiar With Operator's Manual

Read and follow procedures as outlined in the operator's manual. By being familiar with the operating features of a tractor, the operator will develop confidence when the tractor is driven under adverse conditions. Learn the location and purpose of all of the guages and controls as well as other indicators. Knowing where the controls are by memory can allow you to react more quickly in an emergency situation. There have been accident situations where individuals have become entangled in machinery or the power takeoff shaft and rescuers or family did not know how to disengage the equipment. Family members should be showed how to shut down equipment or disengage the PTO in case of emergenc

CHECK TRACTOR BEFORE OPERATING

Pre-Operational Safety Check for Tractors

Before starting any task, ensuring that your tractor is in optimal working condition is essential for both safety and performance. A pre-operational check is a thorough process that guarantees the tractor is ready for safe operation. Here's a checklist to follow before using the tractor:

1. **Tire Inspection:** Check the tires for proper inflation, ensuring they are free of defects such as cuts, punctures, or excessive wear. Well-maintained tires are crucial for stability and traction, especially on uneven terrain.
2. **Visibility and Comfort:** Ensure the windows are clean for optimal visibility. Adjust the seat to a comfortable position and ensure the seat belt is functioning properly. Proper seating ensures that the operator can maintain control and comfort during operation.
3. **Brakes and Steering:** Verify that the brakes are properly adjusted and responsive. Check the steering for smooth response and ensure there is no unusual resistance that could impact maneuverability.
4. **Mirrors and Lights:** Ensure the rearview mirrors are properly adjusted and that all reflectors, running lights, and the slow-moving vehicle emblem are in place and functional. These are critical for visibility, especially during low-light conditions or on busy roads.
5. **Safety Check Around the Tractor:** Walk around the tractor and any attached implement to check for obstacles like stones, debris, or other items that may be in the way. Ensure there are no bystanders in the area to avoid accidents, and make sure the wheels are not frozen or stuck in the ground. Frozen wheels can cause the tractor to flip backward when power is applied.
6. **Loose Parts or Objects:** Inspect for any loose tools or objects that may be on the tractor, particularly near the brakes, controls, or platform areas. Loose items can cause operational hazards or interfere with the machine's functionality.
7. **Service Walkaround:** Walk around the tractor again to perform a more detailed inspection. Examine the tires for wear and proper inflation, check the power takeoff (PTO) shaft for adequate shielding and ensure the hitch has a proper pin and safety clip. Additionally, look for any signs of liquid leaks under the tractor, such as oil, coolant, or fuel, which could indicate potential issues.

Fluid Levels Check

- **Oil Check:** Remove the dipstick, wipe it clean, and check the oil level. Top up the oil if necessary, ensuring that no dirt enters the engine.
- **Radiator Check:** Slowly remove the radiator cap and check the coolant level. Top up if necessary to prevent the engine from overheating.
- **Air Filter and Pre-Cleaner Check:** Inspect and clean the air pre-cleaner and air filter. Remove any dirt to ensure the engine receives clean air for efficient operation.
- **Fuel Level Check:** Ensure the tractor has enough fuel for the task at hand. It's recommended to fill the tank at the end of each day to avoid delays when starting work.
- **Fire Extinguisher:** Ensure the tractor has an accessible and fully charged fire extinguisher. This is essential in case of a fire during operation or refueling, providing an immediate means of response.

By following this pre-operational safety checklist, you can ensure that your tractor is ready for a safe and efficient day's work, reducing the risks of accidents and machinery failure. Always prioritize safety and take the time to perform thorough checks before operation.

HYDRAULIC SYSTEMS AND SAFETY

Hydraulic systems are vital components in many agricultural and industrial machines, powering a wide range of attachments such as loaders, backhoes, and plows. While hydraulic systems offer great efficiency and power, they also present significant safety risks if not handled with caution. These systems operate under high pressure, and the dangers are often not immediately visible, making safety precautions essential.

Precautions Before Servicing Hydraulic Systems

- **Shut Off the Engine:** Always turn off the engine that powers the hydraulic pump before servicing the system. This ensures that the hydraulic system is no longer pressurized and reduces the risk of accidental movement.
- **Lower the Implement to the Ground:** If working on an implement connected to the hydraulic system, ensure that it is lowered completely to the ground. This step minimizes the risk of unexpected movement and prevents the implement from becoming a hazard during maintenance.
- **Relieve Pressure:** Hydraulic systems are under intense pressure, which can be dangerous if released suddenly. Move the hydraulic control lever back and forth to relieve any residual pressure in the system before performing any repairs or adjustments.
- **Support Raised Implements:** If you are working around or under a raised implement, always ensure that it is securely supported by something other than the hydraulic cylinder. The cylinder itself can fail or lose pressure, causing the implement to fall unexpectedly. Use appropriate blocks or mechanical supports to hold the implement in place.

Identifying Leaks and Preventing Injury

Always use a piece of cardboard or wood to check for leaks in hydraulic hoses and fittings. Never use your hands directly to detect leaks, as even a small amount of pressurized hydraulic fluid can be injected into your skin. This can lead to severe health complications, including tissue damage. If hydraulic fluid penetrates the skin, seek immediate medical attention..

Bibliography

Agri-Engineering Today (2021). Role of AI in Predictive Maintenance of Agricultural Machinery.

Agricultural Machinery Manufacturers' Association of India (2021). Technical Handbook on Tractor Maintenance.

Agricultural Mechanization in India: Policy and Prospects (2020). National Agricultural Research System (NARS).

Agricultural Skill Council of India (ASCI) (2021). Training Modules for Tractor Mechanics and Repair Technicians.

AgriTech India (2020). "Breakdown Analysis and Preventive Maintenance of Tractors."

Ahmed, S. (2022). Smart Maintenance Techniques for Agricultural Machinery in India. PhD Thesis, Aligarh Muslim University.

Bharat AgroTech (2021). Agricultural Equipment Usage and Maintenance Handbook for Tractors and Tillers.

Central Institute of Post-Harvest Engineering & Technology (CIPHET) (2020). Maintenance of Post-Harvest Equipment for Better Efficiency.

Department of Agriculture & Farmers Welfare (2021). Operational Efficiency of Tractors and Power Tillers in India.

Escorts Agri Machinery (2019). Tractor Service and Maintenance Handbook.

EU-India Research Collaboration (2021). Technology Transfer in Tractor Maintenance and Servicing.

FAO-South Asia Report (2021). Challenges in Mechanization for Small Farmers in India.

FAO India Report (2021). Mechanization and Tractor Maintenance Strategies for Small-Scale Farmers.

Farm Mechanization Today (2021). "Emerging Trends in Tractor and Power Tiller Maintenance." Vol. 35, Issue 4.

Government of India (2021). National Agricultural Mechanization Policy & Guidelines for Tractor and Power Tiller Maintenance.

ICAR-CIAE (2019). Safety and Maintenance of Farm Power Equipment.

ICAR-CIAE (Central Institute of Agricultural Engineering) (2020). Handbook on Tractor and Power Tiller Maintenance.

Indian Agricultural Mechanization Research Institute (IAMRI) (2021). Service Manual for Power Tillers in India.

Indian Agricultural Research Institute (IARI) (2020). Operational Guidelines for Tractor and Power Tiller Maintenance.

Indian Council of Agricultural Research (ICAR) (2021). Tractor and Tiller Mechanization Strategies in India.

Indian Farmers' Research Council (2021). Reducing Downtime in Tractors and Power Tillers through Proper Maintenance.

Indian Farming Magazine (2022). "Best Practices in Power Tiller Servicing and Repair." Vol. 46, Issue 2.

Indian Journal of Agricultural Sciences (2021). "Challenges and Advances in Tractor Repair Technologies." Vol. 92, Issue 3.

Indian Patent 378429 – Innovative Engine Diagnostics for Tractors and Power Tillers.

Indian Society of Agricultural Engineers (ISAE) (2021). Best Practices for Tractor Repair and Servicing in India.

Indian Society of Extension Education (2020). "Farmers' Awareness and Training Needs on Tractor Maintenance." Journal of Rural Studies.

Indian Society of Mechanical Engineers (ISME) (2020). Advancements in Agricultural Engine Design and Maintenance.

IS 12207:2019 – Bureau of Indian Standards (BIS) for Tractor Performance and Maintenance Guidelines.

IS 13539:2017 – Indian Standards for Power Tiller Testing and Maintenance.

Jain, S. C., & Rai, C. R. (2015). Farm Machinery and Equipment: Repair and Maintenance. Standard Publishers, New Delhi.

Krishi Jagran (2021). "Step-by-Step Guide for Tractor Engine Repair and Overhaul."

Krishi Vigyan Kendra (KVK) (2022). Training Manual on Repair and Maintenance of Tractors and Power Tillers.

Mahindra Tractors (2021). User Manual for Tractor Maintenance and Repair. https://www.mahindrausa.com/owner-resources

Mehta, M. L., Verma, S. R., Mishra, S. K., & Pandey, P. K. (2017). Principles of Agricultural Engineering – Repair and Maintenance. Jain Brothers, New Delhi.

Narayana, T. (2015). Power Tiller Technology and Applications. Agrotech Publishing Academy, Udaipur.

Ojha, T.P., and Michael, A.M. (2015). Principles of Agricultural Engineering – Vol. II. Jain Brothers, New Delhi.

Pund, S. (2011). Farm Machinery and Power. Saroj Prakashan, Allahabad.

Radha, T. and Patel, J.P. (2014). Workshop Practice for Farm Machinery. Kalyani Publishers, Ludhiana.

Singh, G. (2006). Maintenance and Workshop Practices. New India Publishing Agency, New Delhi.

Singh, K.K., and Pandey, P. (2012). Farm Machinery and Equipment. Jain Brothers, New Delhi.

Tiwari, P.S., and Jain, R.K. (2017). Farm Machinery: Theory and Maintenance. Scientific Publishers, Jodhpur.

Verma, S.R. (2011). Tractor Systems and Management. Jain Brothers, New Delhi.

Witney, B. (1988). Choosing and Using Farm Machines. Longman Scientific & Technical, UK.

Index